PROJECT SOFT POWER

Learn the Secrets of the Great Project Leaders

PROJECT SOFT POWER

Learn
the SECRETS
of the
GREAT PROJECT LEADERS

A project fable by

Jeremie Averous

Published by Fourth Revolution Publishing, Singapore
A trademark of Fourth Revolution Pte Ltd
23D   Charlton   Lane,   Singapore   539690,
www.FourthRevolutionPublishing.com

In this book, the alternating of gender in grammar is utilized. Any masculine reference shall also apply to females and any feminine reference shall also apply to males.

Illustrations of the book are by Corne + Enroc, Buenos Aires, Argentina (through Elance.com). Editing was done masterfully by Susan Shepherd, California, USA (through Elance.com).

This book has been sponsored by Project Value Delivery Pte Ltd, a consulting company in the field of project management for Large, Complex projects   –   www.ProjectValueDelivery.com   -   contact@ProjectValueDelivery.com

This book is available both in paperback format and Kindle e-format, through all e-bookstores including Amazon, Barnes & Noble, etc.
Contact us for bulk orders or customized editions!

ISBN: **978-981-07-1539-7**
ISBN e-book: **978-981-07-1540-3**

First print - Printed in Malaysia, May 2012 / worldwide availability on all e-bookshops through Print-On-Demand (LightningSource).

To the Gumusut project team

A live demonstration of Project Soft Power

In the Same Collection

1. The Pocket Guide for those Daring Enough to Take Responsibility for Large, Complex Projects

   by Jean-Pierre Capron

2. Project Soft Power, *Learn the Secrets of the Great Project Leaders*

   by Jeremie Averous

And more to come...

Discover the latest publications and more on:

www.ProjectValueDelivery.com

Contents

Foreword

By Babu Surendran, Project Director

The concept of "soft power" is widely misunderstood and seldom applied by Project Managers. Most of today's Project Directors/Managers have come from the "old school of hard knocks" and do not believe that managing "soft issues" can add value. This book has the capacity to change this mindset and add value to projects.

During my career as Project Director, on large Oil and Gas Projects worldwide, the one thing I realized very early in my career, was that the complexity of projects, especially these mega-projects, makes it almost impossible to be the one who knows it all, even if you are the Project Director! And as I always used to say, no human being lives long enough to know everything about everything. As such, we have no choice but to rely on others in our team to contribute specialist knowledge in specific areas. In order to get this done effectively and have a team with ALL members pulling in the same direction towards a common goal, the team has to have a "buzz," be excited, be aligned and be unified. This can only happen if the soft power approach is taken.

I would strongly recommend this book to anybody who is aspiring to move up the ladder in any project organization towards management positions.

I also want to take this opportunity to thank Jeremie for making this book available to the many who believe in "change," when it is for the better.

Babu Surendran

Singapore, February 2012

Introduction

*"You can't build an organization
which is fit for the future
without making it fit for human beings."* – Gary Hamel

Why is this book important to the project management community?

This book was written to fill a wide-open gap, a real chasm in project management literature. It provides future and present project leaders with a simple framework for practicing and learning the real skills they need to thrive in their profession. As such, this book is a must-read for all people specializing in project management — a challenging and often rewarding profession!

Organizations generally choose upcoming project managers knowledgeable about the technical functions of the organization. Their technical abilities, their experience, and their process capabilities are the criteria used when deciding whether to promote them to project managers. The same criteria are used when it comes to selecting the project core team members.

The new project managers and core team members just do more of what they were doing before.

And then... many fail along with their projects, and many others just barely deliver average results.

They fail, in many instances, because the skills that are important as a project manager are not solely technical skills. As Marshall Goldsmith[1] says, *"what got you here won't get you there."* Though your promotion was based on technical abilities, and these will continue to be centrally important to your success as a new project manager, they are not sufficient. On the contrary: very different new skills are needed.

And indeed, the few outstanding project leaders and core project management teams show very different skills and areas of focus. These managers and leaders also often stemmed from technical positions; but they developed these other skills over time, generally by themselves or by observing and modeling their techniques on those of other successful project leaders.

Stop being a Project Manager. Be a Project Leader.

We have carefully observed successful and respected project leaders, and this research has allowed us to codify those skills for the benefit of the reader. These skills are not complicated. Still, they take practice and discipline to master over the years.

Yet many organizations just throw the new project manager into the turmoil of project leadership without any preparation or any hint that now, entirely new skills are important. This effect is reinforced by the numerous project management books and literature which represent project management as a mechanistic set of processes. Of course, processes are important; they are part of our technical knowledge. Yet the skills that really make a difference are not mechanistic process-related skills.

To start with, a project manager does not manage processes. In reality, he leads his team. So, in this book we will use the 'project leader' terminology as a way to cut loose from the conventional, bureaucratic-minded 'project manager' term.

So... what is 'Project Soft Power?'

'Soft power' is a useful concept that was first defined by Joseph Nye in the field of international relations, and that is used frequently in that context. Joseph Nye defined 'soft power' by opposition to the more conventional 'hard power' of sheer military and economic force. 'Soft power' is comprised of influence, persuasion, and the ability to attract. In the field of leadership, Joseph Nye states: *"many leadership skills such as creating a vision, communicating it, attracting and choosing able people, delegating, and forming coalitions depend upon what I call soft power."*

This book uses the terminology of 'Project Soft Power' as a way to distinguish the five key skills identified in successful project managers and leaders from the usual project management approaches which are based on technical and process-related skills, which would be equivalent to 'project hard power.'

In fact, as demonstrated repeatedly, success in project endeavors is not just the result of the mechanistic application of processes and analytical skills. It is the result of momentum created by the leader, enabling the project team to tackle the seemingly impossible, the never-done-before. Such a movement can only be achieved by applying 'Project Soft Power.'

The five key skills of the project leader

Beyond the technical and process skills, personal and inter-personal skills will in fact determine success or failure. The most respected and successful project leaders demonstrate consistently strong personal and inter-personal skills in their daily behavior. Indeed, their contribution is much more reliant upon the application of these skills than on their technical experience or project knowledge.

The intent of this book is to provide a readily usable framework for these skills. After much research and after boiling down numerous observations, five key skills have been identified as being the main contributors to a project's success:

- a central skill: weave the network to engage with stakeholders;
- two personal skills: constant focus and discipline; and an entrepreneurial mindset;
- two inter-personal skills: leverage the team; and act as a people catalyst.

Anyone can develop and practice these skills. They are simple to understand and difficult to implement at the same time, because they require emotional work.

These skills can be summarized under the framework of 'Project Soft Power,' so as to distinguish them from the traditional 'project hard power.'

Why are the five key skills rooted in emotional work?

What is emotional work? Emotional work is the ability to connect with others at a deep emotional level to create one-of-a-kind experiences. Connecting with others at the emotional level is, in turn, not possible without a personal emotional balance and a consistent emotional discipline.

The concept of "emotional intelligence" popularized in the 1990's is directly related to emotional work. The terminology 'emotional work' is preferred here to convey the fact that it is not a particular talent; it takes effort, consistency and persistence to achieve great results in the field of emotional connections.

It is important to realize that today in our world, in particular as it changes with the Fourth Revolution[2], emotional work is increasingly valued above all other types of work, including manual and intellectual work. This is still shocking to a large part of the population, but that's the reality in our world today, as we shift into the Collaborative Age.

While the value of manual work has already been depreciated for a long time in the Industrial Age, beginning in the early years of the Industrial Revolution, intellectual (processing) work was favored and had the most value. The processing power was scarce; the decision-making was slow in hierarchical organizations. The coordination of geographically distant entities was made difficult by scarce and expensive methods of communication. Under these conditions, analytical and process-oriented approaches were favored and valued.

The premium placed on technical expertise and analytical process-oriented approaches is a direct heritage from the Industrial Age. Many people today still believe it has the highest value, which is why the value of emotional work is still not as widely recognized as it should be. Our education system and our society in general are still in many ways built to identify and promote students with the highest intellectual processing capabilities.

Today we are moving into a new Age, the Collaborative Age. It has been brought about by the availability of cheap and reliable long distance communication and plentiful processing capabilities. In this new Age, intellectual processing-type work will become increasingly depreciated. The work that has real value in this new Age is emotional work.

What makes a project leader successful will increasingly be the way he deploys and uses emotional work, and less and less how much he knows about technical and process issues.

In the field of project management, this book is key to discussing and evaluating to what emotional work entails.

Should conventional project management literature be thrown away?

No, it should not be thrown away. It is important both to know about the different fields and issues that the project manager will typically encounter, and to know that there are processes available to cater to many important aspects of project management.

Yet conventional project management literature is very much mechanistic, process oriented, and linear. It tends to make one believe that the perfect application of certain processes in a particular sequence will guarantee success. While studies show that application of these processes do slightly increase the odds of projects avoiding total failure, the project failure rate still remains very high even in organizations that supposedly apply these processes very well.

Actually, our consulting work shows that there are still many organizations that do not apply the basic disciplines and processes of project management, and this does not enable them even to have an accurate overview of the current status of their projects. An immediate improvement, then, is to enforce the application of some sound processes, which form a much-needed basic framework.

Still, in organizations that have implemented consistently and 'by the book' all the relevant project management processes, conventional project management processes appear to be necessary but not sufficient.

In reality, in the end, each project is a human adventure, albeit a temporary adventure with only a few players, and any person experienced in project management knows that success or failure often depends ultimately on the synergy and spirit of a project team.

Projects have been run successfully in organizations that started from scratch with only the most basic processes and tools in place, or which developed tools specifically for the task. The availability of processes, tools and systems is necessary, in particular for large projects, but it is not a guarantee of success. The crucial success-determining factor is how the project leader demonstrates soft power and empowers the team to effectively implement the solutions needed to tackle the project objectives.

Conventional project literature should not be thrown away, but should be supplemented with a warning that what it contains is necessary but not sufficient in enabling a person to be a successful project leader. The practices of 'Project Soft Power' are here to fill in the gap.

Who should read this book?

This book is directly written for project management professionals, and that comprises all men and women who are involved in project teams. Whatever your position, you can lead in your remit; you are a project leader. So, while this book is for experienced project "managers," junior project "managers," and men and women contemplating the career of project "management," it is also for all of those who, at any level, contribute to the effective delivery of the projects with which they are involved.

This book will transform the conventional "manager" into the leader that each of us deserves to be.

Because projects are at the heart of the Collaborative Age's Open, Fluid Organization[2], project leadership skills will quickly become desirable for everybody. While this book is primarily aimed at people who are already engaged in project management practices, almost anybody can learn from the Project Soft Power practices and use them to great effect in his or her work.

The structure of the book

Section I proposes a fable in which people with some experience in project management will recognize a number of quite typical shortfalls and events. The object of this fictional story is to illustrate the impact of the application of Project Soft Power on project execution in a lively way.

In Section II we will examine one by one the different roles of the project leader when applying 'Project Soft Power:'

- the project leader in his role as SPIDER, weaving his network web;
- the project leader in his role as KUNG FU MASTER, practicing deep focus;
- the project leader in his role as ENTREPRENEUR, investing in the long term;
- the project leader in his role as TEAM COACH, unleashing the team's potential;
- the project leader in his role as a PEOPLE CATALYST, revealing each individual's potential.
- Finally, the Project Soft Power model is brought together in a general summary and overview.

Section III contains a simple Project Soft Power self-assessment to allow you to assess your skill level in the five key skills. This simple assessment is also available online at www.ProjectSoftPower.com where it can be viewed via the Internet at your convenience.

Section IV goes into the details of the assessment results and gives you practical, down-to-earth advice on what to do with your Project Soft Power self-assessment results.

Finally, Section V discusses in greater detail the general importance and value of emotional work, which is central to the concept of 'Project Soft Power.'

The progression of the book goes from a more illustrated and practical approach, toward a more theoretical and conceptual one. The reader can choose to approach the subject according to her preferences. The central piece of the book, the description of the five key skills of 'Project Soft Power' (Section II and III), forms the core of the book's message.

Now ... take action!

"It is one thing to study war and another to live the warrior's life." – Telamon of Arcadia, mercenary during the 5th century BC

Developing these 'soft power' skills is difficult because they are rooted in emotional work. It is because they are so difficult to acquire and develop that those few people who successfully master them are rare and extremely valuable.

You, as well as anyone else, can learn to demonstrate these skills consistently. You can become one of these highly respected and successful project leaders. The ambition of this book is to formalize skills that have not been previously shown that way and to show the path to mastering Project Soft Power.

Yet reading this book is not enough. Knowing those five skills is not enough. You might have already developed some talent in a few of these skills, but it will take time and consistent practice to develop them all. You might need to unlearn some ways of doing things in order to apply these practices. Don't under-estimate the investment. Get some help from a coach or from colleagues to help you overcome the difficult barriers and obstacles that are in the way of your emotional work development.

Do it, because it is worth it, both in your life as a leader of projects and in your community and life in general. I fully hope that beyond reading and sharing this book, you will take action, become a true Project Soft Power leader, and help raise the bar of the project management community at the time when project management spreads everywhere in the wake of the Fourth Revolution.

Jeremie Averous

Singapore

April 2012

Let's keep in touch through the website:

www.ProjectSoftPower.com

so that you can share with the community your Project Soft Power stories.

Notes

1: *What Got You Here Won't Get You There*, Marshall Goldmith, Mark Reiter, Hyperion, 2007: a must-read in the field of career development, from one of the most successful executive coaches

2: *The Fourth Revolution, How to Thrive Through the World's Transformation*, Jeremie Averous, Fourth Revolution Publishing, 2011: a description of the current transformations of the world as we move from the Industrial Age into the Collaborative Age.

The Tunnel Competition Fable

One the next page we introduce the main characters for future reference.

From Norland:
- **Ben**, the athletic high-ranking advisor to the Prime Minister (acting as project sponsor)
- **Harry**, the project manager with thirty-five years of experience in tunnel construction
- **Hugh**, the well-groomed consultant / professor in project management
- **Hilary**, the analyst whizz lady / strategist of the project team
- **Hannah**, the project controls manager
- **Harold**, the bulky construction site supervisor

From Oldavia:
- **Albert**, the food-loving high-ranked Presidential advisor (acting as project sponsor); much enamored of his moustache
- **Simon**, the project leader, a dynamic man in his forties
- **Steven**, the grandfatherly deputy project manager and technical expert in tunnels
- **Sally**, the dark, red-haired and unconventional project controls and contracts person
- **Scott**, the unremarkable and soft-spoken coach
- **Sandra**, the dynamic up-and-coming construction site leader

-1- How the world's largest tunnel-digging competition was decided upon as a practical solution to a political problem

For a long time, the governments of Oldavia and Norland had considered whether to build a tunnel below the sea strait that separated the two countries. Visionaries had dreamt of this project for decades, even centuries. Regarding its economics, however, it was something of a gamble. While a tunnel could only promote closer ties between the countries and promote economic development in both countries over the long term, it would be difficult for any one firm to successfully build the tunnel alone. However, the idea of embarking upon such a large construction work proved to be very popular, as it would create jobs at a time when unemployment was high, and so the public embraced the project with enthusiasm.

In both countries, voices increasingly requested the creation of an economical incentive program consisting of governmental funding for large public works. Carried away by the popular acclaim and viewing this announcement as another way to try to increase their ebbing popularity, the President of Oldavia and the Prime Minister of Norland made a public announcement of the program's launch and signed a treaty to implement it. The signing ceremony was a real show and made headlines in news reports throughout the world.

Building the tunnel would mobilize very significant financial resources and technical skills, and engineers all around the world were dazzled by the magnitude of the project, and waited with bated breath for news of its success or failure. While adequate technology was available for the completion of this feat, the sheer size of the project and the projected length of the tunnel would stretch modern technology to the limit, indeed creating a world record.

The following morning, once the ink had dried on the paper and the Champagne bubbles had evaporated, leaving a hint of a headache in their wake, Oldavia's President's and Norland's Prime Minister's aides were sitting across a long table, looking at each other in something of a stunned silence.

"So, what do we do now?" said one.

"We might have needed to think a bit more with regard to how to implement that," said the other with a sigh.

What they did not say, though, is that they were both a bit scared to have to implement such a bold decision, especially one which they had not had time to plan for properly.

A discussion started. One asked, couldn't implementation of this project be postponed until after the next round of elections?

Alas, the other reminded him, the press coverage was such that not starting construction immediately would not be appropriate. And the instructions from their respective leaders were clear: get to work, create jobs, show progress.

A long silence ensued.

Albert, the top diplomatic aid of Oldavia's President, was a small, round, elderly gentleman obviously in love with both fine food and his moustache, finally said: "My country is renowned for its public works engineers. They have been able to construct impressive public infrastructure that inspires pride in our country. Let us ask them to build this tunnel, as they built so many power stations, dams, canals and highways. This should not be much harder. We will start by creating a committee of experts to form a plan."

Obviously the Norlanders and their leader, Ben, a tall, skinny man in his early fifties who was famous for running marathons on six continents and for maintaining a rigorous diet, could not let such a statement float in the air, and he answered straight away:

"Yes, your infrastructure is great, but your inefficiencies, delays and cost overruns are quite infamous. Your committee will take ages to make a move. We can't afford for that to happen on this project; it would be too costly...at least politically. Elections in both of our countries are in two years' time; we need to show that things are moving! Our engineers are famous for their experience and know-how, and have built much-needed infrastructure in the remotest corners of the world. Let them just do what they are used to doing — albeit on a slightly larger scale."

Another long silence ensued. The Oldavians were trying to figure out how to respond to that disrespectful statement. Their reputation was on the line!

"Very well," Albert finally said, fiddling with his moustache. "I have a proposal for you. You Norlanders love competition and bets. I propose that we should have two project teams building the tunnel, one from each country, and starting from opposite ends. The one that has got the furthest when the two teams meet somewhere around the middle of the strait will win. If we win, you will have to invite me to the most expensive restaurant of our capital city. I warn you, I love fine food in quantity, and—" he paused and smiled "—you will have to eat the same food and just as much of it as I do."

"All right," said Ben after a surprised pause. "That might be the most effective way anyway. Putting a bit of competition into it can only make things better and more efficient. We Norlanders love open competition." After a pause, during which he looked appraisingly in the eyes of his friend, he added: "If our team wins, you must shave off your moustache and come as my support teammate at my next IronMan competition. That might involve a bit of outdoor activity and exercise. Isn't that fair?"

Caught off guard by the absurdity of the terms of the bet, Albert first did not know how to react. Finally he exploded in laughter, held his hand out, and said, "All right, old chap, you've caught me. Agreed." Then he grinned and added: "Anyway, we are sure to win!"

The entire roomful of aides and advisors, relieved that such a decision had been reached so quickly, exploded in roars of laughter and mutual congratulations, and then everybody went their separate ways, busy with more important and complex issues. The President's wife was pregnant and the aides needed to decide how to leak the information to the press so that it would elevate the President's popularity rating; the Prime Minister, on the other hand, had a few lingering issues to resolve regarding a past mistress who was being bit too talkative with the press.

Bureaucrats in both countries were immediately charged with preparing a complete dossier justifying this competition, based the latest economic research. They showed and demonstrated that competition in public works construction served as a way to optimize the allocation of public resources and minimize expenditure. Eventually, it even brought a prestigious academic prize in economics to the lead economists of both governments. (It did not trouble anybody that the Oldavian head economist had written recognized papers a few years before on the fact that state dirigisme was the best solution for breakthrough public works.) Journalists hailed this decision as a major breakthrough in the methodology of government spending, and one which could save billions. Developing countries quickly incorporated this new approach into their policies as they do with all the latest trends in management and governance.

Of course, the real prize of the bet was never revealed.

And that is how the wisdom of these two governments decided, after much careful thought and deliberation, to create the world's greatest tunnel construction project competition.

-2- How project leaders were chosen

Norland's bureaucracy was a model of lean, no-nonsense efficiency. Harry, the most seasoned tunnel-building expert worldwide, was a self-employed contractor with a track record involving the construction of dozens of tunnels in countries throughout the world, and had been hired for a fortune. Harry was in his mid-50s; grooves on his face testified to his many hours spent outdoors, while his protruding stomach revealed that he often diluted the worries of the day in a few glasses of beer. Still, he remained energetic.

A few days later, he was facing Ben in his modern office at the top floor of a high tower. Large windows showed a rare sight of the capital city, far above the roofs of most buildings.

"The project is bigger and bolder than the ones I have done, Sir. Still, I don't see any major difficulty," said Harry, beaming with self-confidence. "It is quite straightforward. A matter of changing a few numbers in my spreadsheets and getting the right people and equipment. I have thirty-five years of experience in this. Don't worry for even a minute. Let me hire the best experts in the country and abroad. We should get this done easily."

"Excellent!" replied Ben, patting Harry on the shoulder. That guy and his team might be the most expensive tunnel construction team of the century, but he was already enjoying the thought of having Albert scrambling, out of breath, to bring him energy drinks and food as fast as possible in a cold, remote corner of Patagonia. "Just do it!"

At the same moment, in the capital of Oldavia, in a dark wood-paneled office which had served as the bedroom of a princess centuries before, Albert was listening to the recommendations of the Engineer General of the Government. Who should be the project manager? What should be his profile?

"I don't need to remind you," said Albert, "how important it is, both for the pride of our country and for the reputation of its engineers, that we win this competition." In the few days since the cabinet meeting, he had realized that he had committed on the spot to something he would regret deeply if the worst came to pass. He had woken to check his moustache frequently during the night for fear of having lost it. What remained of his atrophied muscles were hurting at the mere thought of the terrible outcome Ben had promised him. "What should be his profile, his experience?"

"The logical answer would be to seek the most experienced tunnel construction manager we have. I have a few names here." After a pause, the Engineer General added: "That would certainly be the most logical way for us to decide on the project managers of most projects." He paused, then went on more slowly, stroking his chin as he verbalized his thoughts. "The thing is, I have had a look at the project. You know, I have done some tunnel construction earlier in my life. I think it is easy to underestimate the challenge. It looks like any other tunnel, but the sheer size of it makes it a really special project."

Albert stood up from his two-century-old golden desk and paced slowly on the creaky planks of the flooring while twirling one end of his moustache. His gaze fell upon the majestic painting hanging from the wall, which depicted a man who three centuries before had demonstrated uncommon leadership and then had finally declared himself King of the country at the age of 35, in defiance of all conventions. The painted figure gazed fiercely at Albert.

"Don't you think, then, that although such an expert is undoubtedly needed in the team, the project manager should be a leader instead, someone who is able to mobilize the project team to do things that are unprecedented, scary, beyond the ordinary?" he asked, facing the painting, deep in thought. He knew his old buddy Ben. He would certainly go for the straightforward solution: hire the most expert person he could find. This person would just hire his other buddies and apply the same methods he would have applied to all other tunnels. No, no, he thought. To beat Ben, the only chance was to try another method, and to find someone that could devise unconventional ways to beat the Norlanders.

The Engineer General was clever enough to understand what Albert meant. Actually he was pleased by Albert's unconventional statement, although it would be difficult to sell to his colleagues in the Public Works Council. For the boldest projects, he had always valued leadership skills over age and experience. That's how he had made a name for himself in his younger years. He wished for a moment he could be younger, with fewer responsibilities, so that he himself could take up the challenge. He took a deep breath and could not help but notice how the air was full of a deep smell of wax from history's patina; just the right atmosphere for the historic decision at hand.

"In that case, I have a name to propose. His name is Simon. He is quite young, in his early 40s, and his experience is mainly with highways, dams and bridge construction. He does not know much about tunnels, but he has constantly demonstrated an uncommon knack for leadership that has been recognized by many of his peers. He has sometimes employed unconventional and surprising methods, but he has realized projects that many people thought could not be done. And he has generally delivered results well within the project's planned duration."

After a pause, the Engineer General added: "For me he would be ideal, but can you take the responsibility of placing someone with no experience whatsoever in tunnel construction in charge of such a project?"

Albert turned slowly around, smiling at the Engineer General. He had made his decision. "Assign the best tunnel expert you have as his deputy. We need a project leader, not an expert."

When the nomination of Simon was made public, Ben could not resist a smile. These Oldavians always put in charge young bright people full of energy and leadership, he thought, but they don't understand how technical experience is critical and can make all the difference. What a cruel illusion!

-3- Setting up the project: the perfect handbook start of Norland's team

The teams went to work. The experts, engineers and support teams that were mobilized were amongst the most experienced in both countries, and even worldwide. The project had popularity and appeal. Many had hoped to get on board and were disappointed not to be chosen.

The international press was following the project intently. Word of the competition had leaked out, and many observers were commenting on the moves made by both sides.

Harry, Norland's project manager, had gone on the worldwide market and had hired all the most seasoned tunnel construction crews he could find. That was expensive, but money was not the main concern here. The project team set up its headquarters in a neat office building in the downtown area of Norland's capital city. Harry was an acclaimed technical expert in tunnel construction. He voluntarily recognized that he might be lacking the project management skills to deal with such a large project. He hence hired Hugh, a recognized project management process consultant and expert. He also wanted to be sure to make all decisions based on sound data analysis, so he hired Hilary, a young, brilliant analytical mind, and put her in charge of all the analytical work. Finally, upon a recommendation from a friend, he also took placed Hannah, a young lady who had some experience in project controls, in charge of all the financial and contractual aspects of the project. Harry knew that he, Hugh, Hilary, and Hannah would work very closely together as the core team of the project.

They all met a few days later in the main meeting room of the project team's office. Hugh was a tall, well-groomed gentleman who always wore a tie and a jacket, which was always in stark contrast to the casual look of the rest of the construction team. His eyes gleamed with intelligence behind thick glasses, dissecting people and situations analytically. As a recognized expert in project management processes, he had almost all the available project management certifications available, and was relentless in teaching project management to generations of students. He came to the meeting with his preferred project management handbook, a well-used, thick book, which he placed in front of him, making sure that it was exactly parallel to the edges of the table. Hilary, a plump woman in her late 30s, had already opened her laptop while Hannah, an attractive blonde woman, had just brought her notebook.

"So, how do we set up this team and this project?" Harry said, launching the discussion, after having taken the time to seat himself comfortably.

"Well, that's quite straightforward," responded Hugh immediately, taking his glasses off to emphasize his point. "There are quite a few recognized good practices. We need to define the scope of the work, carefully design an organization chart, make sure that everyone has a clear definition of his job and his responsibilities, split the work into smaller tasks, and plan all of this in an optimized schedule".

The next few weeks were devoted to carefully crafting the project plan and organization chart. Detailed organization charts were produced and communicated, carefully devised with Hugh according to recognized best practices to maximize the efficiency of the team.

Precise job descriptions were issued to cover all the identified positions of the project, and newcomers had to sign them off after carefully reading them.

Workshops were held periodically to review the project team's construction philosophy. Many project schedules from past tunnel construction projects had been brought by the project members. Hilary compiled all this information together in a single schedule that represented the best—and quickest— practices of those projects, scaling resources and supplies as needed to fit the scale of the new project with those analytical methods she held the secret to.

Hannah set up the project controls department. Detailed financial procedures were put in place. The budget and the schedule were elaborated upon with extraordinary precision, thanks to the data available from previous projects—more than 15,000 activities and 2,000 budget lines. Information streams were set up to ensure that data would flow back to the project management office to feed into the progress and cost data that were being collected.

When they met again after this three-month organizing spree, Harry could rub his hands, satisfied: "When I look at this organization chart and execution plan, I feel that we have devised a winning machine to. I feel like we have adjusted all the components with precision and now we just need to get it rolling."

Hugh also looked satisfied. He was proud of the result—the most optimized project setup ever. He would use it as an illustration for his next paper and as a case study that would highlight many of his theories.

At that stage, the Tunnel Engineer Monthly, a specialized magazine, reported, in an interview with Harry: "We have hired all of the best tunnel construction people in the world, and we are now setting up the best team ever in order to achieve this project. We just have only started. The Oldavians have not even been able to find a project manager that has dug a tunnel in his life.

We will reach Oldavia when the Oldavians have not even started digging!" The journalist commented "One can but agree with this statement after looking at the impressive pedigree of all of Harry's project team and the yards of charts and planning on the walls of the project office."

Albert's mood went right down to rock bottom. He even considered whether he should start exercising—a word which had been, until now, firmly outside his experience.

-4- The unconventional start of the Oldavians

Simon, a tall man with a striking deep voice and unkempt long auburn hair, had been surprised by his nomination. He had heard about the competition but could never have dreamed that he would be in charge! A meeting with Albert had convinced him that he had support from the highest level and was quite free to organize his team as he wanted— "as long as you get further than these freaking Norlanders." Albert was realizing how miserable he would get if Ben had his way!

Obviously, Simon needed a tunnel construction expert. Steven joined as his deputy; a short, friendly-looking grandfather-like man in his 60s who was the most qualified Oldavian, with a lot of good international experience and great recognition by his peers worldwide for his technical prowess. Sally, a dynamic and smiling woman in her 40s with short hair dyed deep red, was his project controls and contracts person. They had worked together on his previous project and he was confident that Sally, with her hands-on, field-based approach, would do a marvelous job in avoiding any surprises from that side.

In addition to the people in the team, Simon was used to relying on the services of Scott, a coach and consultant specializing in team organization and effectiveness. Scott, a discreet, self-effacing man without any particularly noticeable physical features, had had a very diverse career before he settled into his career as a coach and consultant. Simon found in his candid external views and his unconventional questions a welcome way to think outside the box and to challenge himself and his team to perform at higher levels. Still, sometimes Scott came out with questions and remarks that seemed as if they came from another world, and it often took some time before Simon could really understand their deep meaning. In the end, over the years, Simon had always recognized how much Scott's challenges had made him improve. While Scott generally did not intervene often, he had allowed Simon to escape more than once from what he had first mistakenly identified as a dead end in his project or in his career.

Simon would have liked to have Scott on the team full-time, to look after the project organization's dynamics. After much hesitation—Scott liked Simon very much—Scott had refused, explaining to Simon that he would have liked to do so, but that he felt he would be much more effective for him and for the project team if he remained as an outside observer, able to say things in an independent way. Simon was a bit disappointed, but got comfort from the promise that Scott would be around to consult whenever he needed him.

They all met together in an excellent restaurant in a part of the capital city that had been built during the Middle Ages, invited by Albert who was curious— and anxious—to meet the team that was to beat the efficient Norlanders. Albert would not talk business before coffee, which allowed them only a small window of opportunity before digestion would set in.

"So, Simon, how are you going to make the Norlanders look ridiculous?" asked Albert, puffing on his cigar, still enjoying the aftertaste of the excellent food.

Simon looked at him seriously and said, "I don't know!" After a few moments of silence, he smiled and added, "But we'll soon find out!"

This statement was not enough to make Albert feel any less uneasy. He left his cigar hanging. He looked obviously worried: "What do you mean, exactly?"

"Well, it looks like it is a project on a scale that has not been done, with specific challenges that need to be addressed. We can't just reproduce methods that worked for smaller tunnel construction projects in the past. First, we need to get thinking about the particular constraints that might impede our progress, and plan the project around those constraints. To achieve that, we need to get input from a network of knowledgeable people. Building this network is the first thing we need to do."

This seemed quite reasonable to Albert, a seasoned diplomat who had one of the best international networks amongst his peers, and with the pleasant feeling of digestion setting in, he almost forgot about his nightmares where he pictured himself out of breath and running a marathon. "Well," Albert concluded, "don't hesitate to ask me for any introductions, and keep me updated regularly. And remember, we need to beat the Norlanders! It is a national priority!"

-5- The power of networking

In the next few weeks, while Sally, the project controls lady, got busy organizing a project office at the location where the tunnel was due to start, Simon and Steven traveled extensively to visit a large number of people and seek their advice. They met all sorts of tunnel construction experts, university professors, and government officials, as well as local authorities, major suppliers, and environmentalists. Simon was a very entertaining and likeable person who knew how to quickly gain the confidence of the people he met. His listening skills were impeccable and allowed him to learn a great deal of interesting information.

In the meantime, the Oldavian team also set up a website that would allow them to easily communicate regarding the progress of the work, and also to allow anybody to input their thoughts about the project in a convenient virtual location. A competition was even set up to suggest the best way to build it. When Harry learned about the website, he exploded in laughter: "Maybe they expect us to give them some advice about how to build a tunnel!"

After one month of meeting with the relevant interlocutors, Simon and Steven were tired, resting in the lounge of a remote airport. They had just visited the world's largest supplier of tunnel-boring machines. Unusually sipping a fragrant glass of red Bordeaux wine, Simon was deep in thought, letting his fingers run through his long hair.

"Steven, does that mean we have a problem? We cannot use a standard tunnel-boring machine because of the site's geology and the diameter of the tunnel? It needs to be specifically customized for us, and that will take time?"

"I'm afraid I understood the same thing as you," answered Steven. "That was pretty clear from the manufacturer. From what we've seen so far, this seems to be the largest stumbling block for the progress of the project."

Simon admired for a while the take off of a heavily-Boeing on the main runway. The evening light was perfect. The fruity smell of the wine added to the perfection of the moment.

"Well, we now know what our first priority should be. We need to put all our effort into securing this machine as soon as possible."

Steven acquiesced. He had already mobilized the engineering and procurement team. They were on the case, and had rescheduled lower-priority work to be completed on a later date.

Two months later, in the Norlanders' project office:

"What do you mean we have a problem getting a tunnel-boring machine?" The sentence rolled like thunder through the entire project office, leading everybody to raise his head.

Harry's temper had just exploded upon receiving the news.

"Why did we not check that before?" he added, speaking more quietly, and turned to Hilary, his whizz analyst.

Hilary looked contrite but shuffled through her PowerPoints to find the right slides: "Well... all the feedback from the previous projects indicated that this had never been an issue. We had time to look into that; we were even in advance of our plan."

Harry's face turned red. "But anybody knows that our tunnel site's geology is particular!" he erupted.

Hilary and Hugh, the project management specialist, looked at each other. They had not known.

Harry took a few minutes to calm down and sit back on his chair. "So, what do we do now?" he said in a low tone.

Hilary jumped in: "Well, the supplier said he would take six months to build the right machine." After a short silence, she added in a low, slow voice "He also said it will be finished relatively quickly because he has already done the engineering for the Oldavians. They ordered their machine two months ago."

They thought they were witnessing Harry having a heart attack.

-6- The renegades' team

Simon had brought his team together for their first town hall meeting.

Sally, who was officially in charge of project controls and contracts, and unofficially his right-hand support regarding all the internal operations of the project team, had coordinated the recruitment. She had a marked tendency to look for unconventional people, whom she called, half-jokingly, her "renegades." While she undeniably had a preference for red-heads, her overall logic for doing this was extremely clear. She was looking for people that had been successful even though they had not started with connections they could benefit from, and thus, were people that had an intrinsic habit of hard work and persistence in the face of adversity. She looked for people that did not stick to the conventional way of doing things and were always striving to find unconventional solutions. Many of them were people that did not necessarily ask for permission before proceeding. Many of the recruits were immigrants or children of immigrants, or were people who had built a solid reputation for themselves as professionals while stemming from socially disadvantaged groups.

When Simon first met the team, he really thought that this time, Sally had gone a bit too far. While most had worked in construction projects, only a few had built tunnels. Still, he felt an energy level among the team members that made him feel good. They all glanced at him, proud to have been chosen for this milestone project, which could mean so much for their future careers.

Simon's speech was short and to the point. This was a hallmark project, and failure was not an option. Safety was paramount. He counted on the team; although everybody had got an assignment to start with, he made extremely clear that he would expect people to change assignments according to the project's needs, their talents and interests, taking roles where they could bring the most to the project. He also made clear that he would retrench people who would not perform.

When Simon opened the floor for questions, a tall, tanned, muscular brunette took advantage of the opportunity to speak. "My name is Sandra. I'm here as a surveyor. Simon, many of us have a question. How can we expect to win this competition? The Norlanders have the best names that can be assembled on the face of Earth, while the press just makes jokes at our inexperience."

Simon was genuinely pleased this question was asked. He smiled. "Thank you, Sandra, for this question. I know you are all wondering about the same thing. My response is short and simple. Do you watch football? Which team gets the best results: the team that hires expensive stars and tries to get them to play together as a team, or the team who hires less known players but manages to have them play as a single, performing team?" After a pause during which he looked over the room at all teammates, he continued: "We're going to be the best-performing team around, and for that I need your help. When you see something that needs improvement, if you believe you can do it, just do it, even if that's not part of your job description. We will win or lose as a team."

The atmosphere during the drinks that followed the town hall was electrifying. Everybody had internalized the challenge. Simon spent hours answering pressing questions and requests. Albert, who had come to witness that first team meeting, was impressed. He almost felt too conventional. At least he could boast of his moustache, he thought, stroking it carefully. That is...until the end of the contest.

-7- The midway diplomatic dinner

"So, Ben," said Albert while carefully wiping the corner of his lips with his towel, artfully avoiding his moustache, "are you getting ready to offer me this sumptuous dinner, and share it with me?"

Ben was not in a joking mood. He had called Harry the day before, for the first time since the start of the project, and Harry had painfully explained that although they were sure to win, there had been a few occurrences where the Oldavians had decidedly been better performing. In particular, they would probably get the tunnel-boring machine earlier. Ben could not believe what he heard: the Oldavians would get the boring machine earlier! The explanations had been somewhat confused, but Ben was experienced enough to understand that the superiority which he believed he could boast was not proving to be entirely obvious—probably on the contrary.

Seeing that his friend was not ready to take this issue lightly, Albert pushed back his plate and looked Ben in the eyes. "Ben, we are both diplomats. You know about Joseph Nye's 'soft power' concept. In the field of international relations, it might be cheaper and more effective on the long term to deploy 'soft power' influence rather than try to solve issues using the traditional 'hard power'—using the military or embargos."

"Yes, of course," Ben answered, crossing his arms and looking bored. He had just come out of a long meeting with the Prime Minister who had promoted soft power rather forcefully—a nice way, in Ben's opinion, to justify cutting the defense budget to reduce spending. "Just another trendy concept," he thought.

"Well," continued Albert, "I believe that our contest is a kind of an experiment that will highlight the differences between soft and hard power."

"How so?" Ben looked suddenly interested. He knew that under his distant air, Albert was a very sharp analyst of situations. Actually, Ben would not admit to himself that he had a lot of admiration for Albert's encyclopedic knowledge and constantly clear vision of complex situations. Albert was definitely much more clever and observing than what he looked like at first glance.

"Well, our project teams have two totally different way of approaching things. From what I infer from the press, your project manager applies project management processes by the books, and he leads by the plan, force and fear. He relies on proven processes. On our side, our project leader has a much more flexible approach, focusing on the constraints of the project, but mainly, relying on teamwork and on using the talents of his people to fullest effect. I increasingly think our competition is a small experiment pitting hard power against soft power."

Ben looked at his friend for a while, thoughtful. This difference had not occurred to him. But then, Ben had spent a lifetime in bureaucracy. He was apprehensive of situations where roles and objectives were not defined properly. He just loved hard power. Soft power was too ... soft for him.

With a twinkle in his eyes, Albert continued. "Of course, I am not saying that the Oldavians kept such a level of civilization that it allows much easier soft power to be deployed, and that Norland's permanent efficiency drive just killed it—and any remaining hint of proper civilization, like fine food and enough time for lunch. When are you going to start eating some decent food again in Norland? Anyway, isn't it an interesting experiment, my friend?"

As they shook hands and parted, Ben looked at Albert walking away down the street. Soft power. That was quite an interesting concept. But soon he shook his head. What softy nonsense! Processes and efficiency were the most crucial factors in determining a project's success. The Norlanders would win. Easily. They were always the most organized and efficient.

-8- Two opposite leadership styles

Harry's and Simon's working styles were as opposed as the running styles of a hare and a snail.

Harry always followed his electronic dashboard in detail and went very deep into all the detailed activities of the project. His scathing emails about late deliverables came quickly after any missed due dates and were soon feared by the team. This progressively drove team members to put a lot of margin into their estimates for the duration of tasks.

At the same time, Harry responded willingly to the many solicitations from journalists, conference organizers and other stakeholders. At one point he was absent from the office for weeks in order to attend a remote conference where he had been chosen as the keynote speaker. During that time, the team relaxed and breathed sighs of relief, as Hilary's reminders were not as respected as Harry's had been, coming as they did from an analytical genius who did not know anything about tunnel building.

Whenever something happened on site, Harry would ask for all the details and hold the local supervisors for hours on the phone trying to understand what exactly had broken down, and why, and how to remedy the situation. He was a real nuts-and-bolts manager, feared by the team for his attention to details. His inclination to peruse the drawings and electrical wiring diagrams of the tunnel-drilling machine was well known. He aimed to know the machine's engineering better than the people that would operate it. Soon, he knew by heart the name, location and function of all the components.

As they were assembling the tunnel-boring machine on site, two components did not match perfectly. The problem was not immediately obvious, though, and Harry called in a long meeting with representatives of the manufacturer to find the cause of the problem. Tons of diagrams and drawings were studied; 3-D models were developed so they might be projected on the screen and the results closely examined. After long hours, late in the night, a discrepancy was found. This 'aha' moment did not last long, however, because it was quickly discovered that this particular discrepancy was not related to the original issue. The search for the solution lasted another few days, with Harry being intensely involved. He was so focused on the problem that he was temporarily oblivious of the rest of the project. The problem was that at this very moment, setting up the tunnel-boring machine was not the critical task of the project; the most important task at the moment was boring the entry shaft. As Harry, and therefore his management team, focused the majority of their time and attention on solving the machine components problem, precious days were lost in digging the shaft.

Simon's style was the complete opposite. He had started his career exactly like Harry, though. He was an engineer by training, and had been passionate about all things technical. As his responsibilities had grown, as well as the size of the teams he led, he still continued to delve deep into all the technical issues that occurred in his various projects. This made him stay later and later in the office. He had even started counting the emails he sent out each day. His personal record was 143 messages sent in 24 hours. Wow.

And in the rare moments he was at home, his mind would constantly wander back to the projects' details and other issues. He had a family by then, and his wife increasingly began to frown at this invasion of home life by his work.

It was about at this moment that Simon had met Scott, the coach. That was not by chance—his boss had requested for him to meet Scott, after having told him that he had a problem he needed to sort out. "Which problem?" Simon had wondered, as he considered himself very successful, in full control of his circumstances. His boss had answered, with a fatherly smile: "Look, you're an excellent guy, but it is a pity not to see you rising into your new responsibilities as you ought to be doing. No drama; we all need help at some stage in our careers. I just hope you make the best of it." Simon was perplexed. He did not really feel he had a problem.

The day Scott entered his office, Simon was not impressed. Scott was an average guy. He was of average size, in good but not particularly noticeable clothes, clean-shaven, had short hair, no particularly unusual physical features—the type of guy you could pass in the street without ever noticing him. Still, as they started talking, Simon noticed how intensely Scott was listening, solely concentrated on what he was saying. Scott's gray eyes were focused on Simon. Soon, Scott had established a connection.

Almost conversationally, Scott asked Simon how things were in his life, with his family. Confident, Simon explained that he was a bit overloaded, which was a normal consequence of his increased responsibilities, but otherwise things were fine. He started thinking that if all his boss had found out was that he ought to have a nice chat with this guy, his issue must not be that problematic.

It was at that moment that Scott asked the first of his outer-space questions, right out of the blue: "Simon, how effective do you think you are as a project manager?" Without giving the matter much thought, Simon described how efficiently and diligently he was answering emails, how passionate he was about chasing technical issues, how he liked dealing with his team's issues.

A long silence ensued. As Scott was still maintaining eye contact, Simon became uneasy. Scott finally said: "This is not what I asked." Pause. "There is a difference, a wide gap, between being efficient and being effective. Just now you described to me how efficient you are. You did not explain to me how you were effective in reaching the objectives of the project."

Simon was stunned. It was the first time since high school that someone had questioned the fact that being busy, productive and efficient was maybe not what was expected. He did not know how to respond. How important for the project was it that he was known for answering all emails within the same hour? Under Scott's piercing stare, Simon was suddenly not so sure anymore.

Still staring at him like he would at a rabbit frozen in the beam of a flashlight, Scott continued: "For me, you could stare at the ceiling for the entire day, doing nothing else, and still come up with the one great decision that will make the project happen. That's what I call being effective. Now let me ask you again: how effective are you as a project manager, reaching the objectives of the project?"

That night, when he drove home more than an hour earlier than usual, Simon was frowning. This Scott guy had just questioned all of Simon's education, social references and habits. Deep inside, however, Simon knew he was right. At the end of the day, at the end of the project, people would remember the project's outcome, not how many emails Simon had written that day, or what his personal record in email sending was.

In the following weeks, Simon met with Scott on several occasions. They discussed more deeply what an effective project leader was, not in theory, but in practice. Scott would only ask questions, making Simon struggle for the answer, sometimes for several days.

Scott also provided feedback. He organized a thorough leadership assessment of Simon, which showed a consistent picture of his strengths and weaknesses: a great working power, but a need for more vision and drive and better prioritization. Scott also did one thing that completely transformed Simon's life: he asked Simon's family for feedback on how Simon was doing.

Much to Simon's surprise, the feedback was that Simon was increasingly distant, and by keeping his mind constantly preoccupied with his work, he was not really present or attentive at home. This particular discovery gave Simon the drive to change.

Simon was now known to be a leader. He practiced listening to his team, both in the office and in the field. He did not interfere with the way team members tackled issues if they did so effectively. At the same time, he did not let himself become swamped by technical minutia or the details of other pending issues unless in those cases where his input could be worthwhile. He was now able to disconnect his workload from his time spent at home and to better prioritize events at work. This had also allowed him to broaden his perspective, and thus, caused him to relate better to a number of stakeholder concerns he could have easily overlooked before.

-9- Simon's Rules of Two Plus One

Scott gave Simon a few books to read, amongst which Eliyahu Goldratt's <u>The Goal</u>, a famous book about the Theory of Constraints, featured prominently. The book's action took place in a manufacturing environment; Scott assured him that the reasoning was that at any given time, there are only one or two constraints driving the performance of the system, and this reasoning also held true in project environments.

Finally, much to his amazement, Simon discovered that being effective was not:
- being on the everybody's back all the time;

- letting everybody know of all the issues encountered in the project;
- involving himself personally in all of these issues, or expecting himself to solve all issues;
- being too analytical and developing a detailed overview of the entire project.

Being effective was rather about:
- identifying the one or two main drivers or constraints of the project
- focusing solely on these drivers
- subordinating everything else to these drivers

And after a while, by testing these general observations and practicing effective leadership in accordance with these points, Simon came up with what he called the Rules of Two Plus One.

The first rule of Two Plus One is about focus. At any time, as a project leader, he would focus only on two items: one short-term priority and one medium-term priority. The short-term priority would change every one to two months as the issue was solved, and the medium-term changed approximately every six months as it in turn was solved. In addition to focusing on two priorities, he would permanently focus on the project team's health and spirit. Hence, the Rule of Two Plus One.

Not only would Simon follow this rule, but he would let his team know very publicly what his priorities were. After some time, he even decided to post them outside his office.

As they joined, his project team was initially surprised to see, hanging outside the office, a sign that read:

Two Plus One Focus

Our focus is currently on

1) deciding the entry shaft strategy

2) getting the right tunnel-boring machine on time

And as always, promote a healthy team spirit: foster engagement

Sally, his right arm in running the office, would interpret the sign for the astonished newcomers: "Simon's philosophy is that in order to be an effective project leader, he has to focus on what is the most important for the project—and not change his priorities too often." She would also add, "and the worst is that he really sticks to it—he is very disciplined. He will listen to the other issues that are happening, but as long as they are not as important for the project outcome, he will not focus on them. He will expect us to deal with them and move forward. Now, don't stand in the way of his priorities! He will really push you out of the way, literally. That's why he wants you to know what they are, at any given time."

And indeed, Simon was now very wary of how he was spending his time. As a project leader he had obligations, of course. He could not just focus all of his available time on his chosen priorities. But still, he always managed somehow to expound on these priorities at every occasion: with bosses, with stakeholders, with the press, with the members of his team. And his messages were always aligned with his chosen priorities. As everybody focused on them, these priorities tended to be solved much more quickly and could then be replaced after they had been fully resolved by the new constraints for the effective delivery of the project.

Simon had another rule of Two Plus One. This rule was about discipline. Not discipline in the sense of micro-managing the people in the team, or maintaining discipline in the sense it is used in a school yard. It was a rule of personal discipline, extended to the project team, regarding a few key issues that were vital to monitoring the project. Simon was very flexible, but his team knew that whatever else happened, he expected them to be disciplined about those very few things.

Simon's discipline rule of Two Plus One was:
- ensure the back-office processes work reliably
- focus on the priorities
- and, safety above all—because no project is worth a human life.

In particular, he was known to expect extremely regular project reviews and he perused the project reports in detail. As something of a creature of habit, his team knew he would never miss a monthly update and subsequent discussion. And he always supported his team when they needed help or additional resources in order to get the usual project management processes up and running.

Still, contrary to Harry, who spent most of his time looking at indicators and at screen after screen of systems output, Simon made sure to spend only a limited portion of his time on the internal workings of the systems. He was confident that Sally would highlight any looming issues early enough that he would be able to address them before they became a serious problem, and that enabled him to devote most of his time to other priorities.

-10- How the two project teams came up with two different methods for shaft optimization

While the team waited for the Beast (the nickname given to the gigantic boring machine), one major task before they could start digging the tunnel was to prepare the site of the tunnel entrance. This was critical because of the sheer volume of soil that would have to be extracted from the tunnel and disposed of, and the logistics that were needed to keep the Beasts running day and night were complex and multifaceted. The site of the tunnel entrance was a large piece of land that had to be cleared, leveled and prepared with all the necessary equipment before the Beast ever arrived to begin digging. Also, a large shaft was needed to start the tunnel directly in the right geological layer. Digging this shaft was a significant investment in time and resources. The diameter of this shaft was a major project decision: if the diameter of the shaft were larger, it would make lowering the Beast easier, as well as improve the overall logistics during the boring phase. If the diameter were smaller, it would cost much less time and money.

Both teams went through the process of deciding the diameter of the shaft.

At the Norlanders' office, Hilary, the analyst-cum-data cruncher, had designed an optimization model with the help of Hannah, the project controls manager, using time and cost data from previous tunnel construction projects. She had spent entire nights chain smoking and running an impressive number of simulations to find the optimal diameter. She finally raised her eyes from her screen and declared to the expectant Harry: "The optimal diameter of the shaft is 20.345 meters. The model has got 267 variables so it took me some time to put it together, but it's running fine now, and that is the correct number." A few tunnel experts in the room were looking at her in disbelief: in all the projects they had done previously the diameter had been at least 25, if not 30 meters! As they started arguing, Hilary responded with an air of certainty: "It's not my fault if you are used to digging tunnels following your century-old rules of thumb instead of relying on modern non-linear optimization techniques! Using this diameter will save millions in operation costs and dozens of days on the project."

Harry had already spent hours with Hilary looking at all the details of the computer model. He did not understand everything, but it seemed to him that all possible parameters had been considered. It could not be wrong; the best people had contributed to it. Soon the decision was made: the shaft would have a diameter of 20.345 meters, and not a centimeter more. Anyway, this would allow the team to dig much quicker and start boring the tunnel before the Oldavians, almost certainly catching up with them, which was good news. The work on the dig site began.

On the Oldavians' team, this exact problem had likewise been identified as very critical. Experts could not really agree on what to do, so Simon finally said: "All right, guys, let's build a physical model and see whether all this will fit in and operate." After wondering whether this was a joke in the era of computers and iPads, some of the team members who were hobby-modelers volunteered, and a scale model of the shaft, the tunnel-boring machine and all its components was produced. Some desks were moved and the large model, waist high, was built with transparent materials so that everybody could see how everything would fit.

In addition, Simon required the team to consider some scenarios where things would not go as expected: where the tunnel-boring machine would need to be taken out and overhauled, or where some major parts would have to be replaced. These scenarios were also played in the model, with the team grouped around it. This way, they all began to understand how all this would operate.

Simon liked playing around with the model; he felt like a small boy playing with model trains. The model of the Beast was inserted and removed dozens of time in different ways.

Finally the decision was made. The best shaft was not round, but elliptic, with a longer dimension of 35 meters and a smaller dimension of 25 meters. That might take longer to build, but Simon and the team knew that by making this investment, they would spare themselves some headaches down the road, particularly in the case of unexpected events. So they went forward with this design.

When the designs of the shafts were released to the press, Harry confidently communicated the fact that, thanks to a proprietary design optimization technique which the Oldavians were quite incapable of doing, his team would be ready to dig the tunnel one month earlier than their team. There was much praise of the cleverness of their solution, and technical articles were written that went on to gain awards in the professional conferences, while all the commentators wondered about the elliptical shape of the Oldavians' shaft. Was it an attempt at incorporating a modern design?

Albert obviously became concerned when he heard about these developments from the news media. He called Simon. Simon just asked him into the project office, and showed him the model. He handed Albert the model of the different components and just explained to him the logic. At first, Albert was stunned. Is that why we pay them millions? he wondered. To make models like ten-years-old children? Then he tried it himself. He soon got into it, and became so engrossed he (almost) forgot about lunch!

Simon finally wrapped up the day by stating in front of Albert and the team: "I don't know much about tunnels, but I know about construction projects. I've never seen any happen without major breakdowns, sudden changes and unplanned events of all sorts." After a pause, he straightened up further and added, looking Albert in the eye: "As the project manager, I prefer to invest in preventive measures now than suffer later in trying to correct a situation I could have easily avoided." The next day, someone printed Simon's last sentence in large characters and hung it on the wall of the office so that everyone could see it.

Invest now in preventive measures rather than suffer later

That was the philosophy that they shared and made them so special. They were proud of Simon. This would only be the first of many quotes to hang on the wall.

-11- Teamwork styles

The Norlanders' team was working according to plan. People who reported to the team were assigned a role, a job description, and tasks, and they were completing them in order, reporting any issues to their manager. That was highly efficient from the first day onwards. The engineers would often eat their homemade lunches at their desks, but generally left for the day at the official time except when they were extremely late on a deliverable. After all, the duration of the work had been properly calibrated.

Whenever a question arose, decisions were quickly made. Hilary would run an analysis of the issue, as complex as the situation required, and then put forward her recommendation to Harry. Because he was confident that the analytical approach was the most reasonable way to choose between options, Harry generally endorsed Hilary's recommendations.

Because of this centralized and somewhat mysterious decision-making process, team members did not really feel in control. They were merely raising issues, answering clarifying questions or finishing more complex analysis tasks, and merely expecting Hilary's computer (the Oracle, as they jokingly called it) to utter the answer. As much as the project team members hated it, they started having fun about it and even placing bets among themselves on the Oracle's outcome for each tricky question.

On the Oldavians' side, progress had been relatively good so far, but they were no farther ahead than the Norlanders, who had managed to stay abreast. Contrary to the Norlanders' team, the Oldavians' team had taken some time to settle down and organize itself. Simon had spent considerable amounts of time and effort in letting people get to know each other, both in the office and in less formal settings. Lunches were taken together in the cafeteria and the managers, Simon included, would sit with different groups of people each day so that they would know everybody.

At the same time, Simon repeatedly emphasized how challenging this project was. This challenge galvanized the team, and many were putting incredible effort and hours into the project. There had already been some departures, though, of people that did not fit in with the team or who did not want to follow the team's chosen rhythm. That was all right; project teams in stressful situations always select the main contributors by themselves, and those who do not contribute are pushed away from the real action by the others.

Sometimes Simon had to step in to accelerate the natural movement and let some people know that they were not required any more. It was important not to let negative, less dedicated people pollute the general dynamics of the team. Still, Simon was always impressed by how challenging projects created an accelerated selection process for the team members. The core team, the main contributors, the pillars of the informal organization of the project, was forming by itself. It was just a matter of maintaining the pressure of the challenge presented to them and accelerating the natural team selection process.

Simon was not deterred by the idea of having difficult discussions with his team members if that was required for success. He was able to give feedback that was critical without being harsh. Everybody knew they would be treated fairly in any case, and they knew that Simon was able to stand by his remarks, as well as being willing to give a second chance to someone who had made a mistake. People that had gone through the process generally remembered this experience as something of a life-changing moment, and gave double the effort to compensate for their inadequacies. Though a few people got discouraged and left, they were not the majority.

Other tough conversations were sometimes needed to convince people that they could take bigger roles than what they had previously thought themselves capable of. One of the toughest conversations Simon had was with Sandra, the surveyor. Sandra exhibited a great potential for leadership, but she thought of herself as "only a surveyor" and did not see herself stepping up to a wider role. It took all of Simon's persuasive skills and several discussions to convince her to try to lead a small team at first, overcoming her fear and her self-imposed professional identity as a surveyor in order to take a leadership role.

Simon had asked Scott, the coach, to support the setup of the team, because Simon knew from experience that it was very important to have it right so as to really be able to unleash the potential of the project team later on. So, Scott had organized periodic "mood and team development surveys" in the form of anonymous electronic surveys, and he was also there in person sometimes, just observing and listening to the team as it was working, making progress or struggling with the tasks at hand.

The team members were at first a bit suspicious of these "mood surveys" and thought that it was another of these feel-good initiatives that management uses so that they look like they care about what people feel. They were pleasantly surprised when Simon made a point to share the results of these surveys and to take action with regard to the issues that were raised!

Scott happened to be there on a day when a meeting was taking place on a highly debated subject. It was about how to organize the storage of the soil that would be excavated from the dig site. Different people had quite opposing opinions about the general layout of the site as well as the size and shape of the storage areas. Both sides of the debate were passionately argued and the discussion even became a bit heated between the supporters of two popular by incompatible options. Simon was both impressed and annoyed by the way the meeting had progressed, and called for a ten-minute break. He needed Scott's advice. Simon was uncomfortable because he knew that somehow, a decision had to be made, but at the same time he did not want to disrupt the positive dynamics of the team.

Simon did not even have to fully explain his issue to Scott before Scott smiled and said: "I'm sure you feel that you are in a tough position, Simon." As he was used to doing, Scott let the silence fill the conversation. Simon moved uneasily on his chair. Scott finally broke the silence. "How soon do you need to make that decision?" Simon answered quickly: "I guess we need to make it now, because the layout and setup of the site will depend on it." Another silence followed. Scott asked: "How can you get the decision made and end up with the team being even more passionate and involved?"

Scott observed in Simon's eyes that he had gotten him thinking. After stroking his chin, Simon slowly smiled and answered: "As usual, Scott, you remind me that instead of placing these objectives in opposition, there are certainly ways to reach them both." After short break he came back and said: "I guess I've found a means of doing that. I just needed to get over my preconceived idea that, as a leader, I need to make the call. But then obviously I am not as competent as many people in the team on that subject. I'll get the entire technical team to vote on the options." He nodded to Scott. "Thanks, Scott. It looks logical in hindsight, but I know that I need you to ask those seemingly innocuous, but in reality fairly challenging, questions of yours."

Two days later, after some intense campaigning by many people for their preferred option, the vote was cast. The decision was made by the technical team. Simon reminded everybody that now that the decision had been made, he wanted everybody to stick positively to it. And after that, he welcomed such heated debates when they arose around key decisions that were necessary for the project.

Indeed, the positive group dynamics of the project team increased tremendously. They were no longer just members of the project team. They were collectively in charge of the project's destiny.

-12- Dealing with the Press... a matter of priorities

The two teams went on to prepare the site, test the equipment, and dig the shafts. Specialized journals and magazines followed the teams' progress intently, showing how the work was going on from month to month, and commenting on the odds of each team ultimately winning the contest.

One thing was sure: the idea of the international competition did bring the media's attention to the matter. Albert and Ben were not so sure that this was a good thing; they would probably have preferred that this particular project stay out of the spotlight. On the other hand, there were a lot of positive comments about how, for once, the two countries' governments had followed through on plans for a high-stakes project, so that made them feel good.

Of course, the task of digging the shaft went more quickly on the Norlanders' dig site, as they had chosen to make it smaller. In addition, Harry was all over the place in magazines and newspapers, highlighting his superior leadership and organization, sending regular pictures of his team's progress to the press, and inviting journalists and stakeholders to visit the site. On the other hand, the journalists were frustrated that the Oldavians did not communicate that much; while they communicated in a minimal way, Simon was said to be always too busy to respond personally to their requests. Of course, the journalists interpreted that as a compensation for the weakness of his leadership style, and assumed it to be evidence of a looming defeat. The particularly virulent tabloids, so widespread in Norland, were having a field day proclaiming the weaknesses of the Oldavians, describing them as "people of this old country hampered by their traditions and entrenched bureaucracy."

Albert did not understand all the technical details. But he was concerned about the negative media exposure and could hardly wait to share his concerns with Simon. He had also found his country's project team to be still a bit disorganized, almost messy. Albert wondered whether the neat and superior-looking organization of the Norlanders' project team, which had been so often praised in the press as a perfectly engineered clock mechanism, would not win the day.

Simon was expecting the call. Actually, he had been expecting it earlier! As he answered, his eyes wandered across the wall, finally coming to rest on the latest Rule of Two Plus One revision he had posted:

Two Plus One Focus

(version 2)

Our focus is currently on

1) ensuring the tunnel-boring machine will be assembled with no hiccups on site

2) planning the excavation team and organization

And as always, continue to promote a healthy team spirit: candid feedback

---

Albert was not in a very good mood that day. He had just had a cheerful call from Ben telling him to reserve the dates of the next year's IronMan in his calendar. So, he inquired directly about whether the task of digging the shaft was being carried out with the right priority level. Albert almost choked when Simon answered: "Albert, this is not in my top priority list at the moment."

A somewhat annoyed silence ensued. Finally, Simon explained: "Look, Albert, I know what the press is saying, but we are not going to go faster even if the shaft is finished tomorrow. What's the use of accelerating this one task if the rest of the gear is not going to be ready? I have good guys looking at the shaft at the moment, and they are doing a reasonable job. As for myself, as a project manager I now need to make sure that we have everything ready to start boring when the shaft is completed at last."

Albert was not really sure he understood the point Simon was making. "So," he replied cautiously, "what you are telling me, in essence, is that you are not trying to finalize the shaft construction as fast as possible, because there is no use burning resources doing that if the next activities will not be ready in time?"

Simon was relieved that Albert was so quick in understanding this counterintuitive approach. Simon added, "That's exactly right, Albert. As I told you before, I focus exclusively on the main stumbling blocks that could prevent my team from reaching the main objectives of the project. That's also why you don't see me defending our progress in the press. What good would it do for the project at this moment? What's the use of me sitting day-in and day-out in radio and TV talk shows? That will come later."

Albert was not sure about Simon's last statement. "You know this is starting to be a problem, and soon I'll have the Opposition clamoring for my head. It'll be chopped off and carried around on a spike if we don't start to show visible progress. Somehow I feel that might be even more problematic than having my moustache shaved off, my friend."

Simon was moved a bit by Albert's concern. He liked Albert's character and certainly did not wish such an ignominious future for him! Still, he had to make his point. "Albert, we have had this discussion about the media strategy already. And if we start doing a lot of stuff now in order to respond to them, we'll be in reactive mode, which will be worse. We said we would over-communicate only when we had begun effectively digging the tunnel. I propose we stick to that. I don't want to spend my time on this, since it will not help the project overall. For the sake of the success of the project, I need to focus on my two priorities, not on the whims of the press."

As he hung up, Albert was really happy they had chosen Simon to lead the team. He clicked on Ben's original email and answered succinctly: "Thanks for the calendar update, Ben. As I'm a very fair player, here is the list of the three restaurants I would consider for the invitation you will owe me..."

-13- How mussels can stop a tunnel's construction

It started like any other day for Robert, the president of his region's Mussels Growing Association. He was sailing on his small, specialized boat, a square aluminum vessel that had served him well for years. With his pipe in his mouth, his tanned and weathered face, and his favorite dark blue mariner's cap, he was a photogenic asset of the coastal town where he lived. Tourists liked to have their pictures taken with him, and in good weather during the tourist season he would earn extra cash by ferrying them around the beautiful, wild coast.

Cash was a bit strained these days. Mussel growing was not as profitable as it has been a few years before, and lately, new health regulations had made it even more difficult to sell them. Last year they had had to throw out half of their production because of some toxic algae that was found in minute quantities in some mussels.

The sky was low and gray, with periodic showers. The small boat was struggling to overcome the swells stirred up by the wind, and as the boat cleaved through the waves the breeze shifted, sending a spray of water across Robert's face. He liked the sour smell of seawater, though, and hardly minded. He could see, high above the cliff, the cranes and the bright lights of the shaft construction site. This project was good, he thought. It had brought increased activity into the sleepy community; a few restaurants had reopened and a number of young people had come back. At least it will last for a while, he thought.

As he reached the areas where his mussels bred and grew—a set of vertical wooden posts in a shallow, protected area—he slowed down his boat, moored it to one of the stakes, and started his usual routine of collecting mussels.

Soon, however, he stopped working, perplexed. All the mussels he was bringing up were dead. His stress level rising, he moved his boat to the other side of the mussel growth site, only to find that his mussels were in bad shape there too. Robert could hardly believe it. In all his life he had not seen anything like this happen. What was wrong?

Then he looked back at the light and heard the noise of the construction site over the breeze. It must be the tunnel. It could only be the tunnel! He sailed back as quickly as possible to the harbor.

A meeting of the local Mussels Growing Association was organized and would take place that same evening. Some mussel growers had found similar issues in their breeding sites, but not to the extent that Robert had; and some others did not report anything wrong with their sites. Still, the mood in the room was tense. These men were discussing the livelihood of their families and that of most of the village. They sat sternly, arms crossed and faces closed, and the silences were heavy with anger.

What could be the cause of this sudden spike in mussel mortality rates? Uncontrolled discharges from the construction site? Vibrations? In any case, something had to be done about the problem, and fast. A decision was quickly made. Robert would call the site management and also contact the local representative in parliament to get his political backing, just in case. And he would also call his colleague and friend Alan from the other side of the sea strait to see whether they were having the same problem.

Since the beginning of his assignment, Simon had continued weaving his network intensively, showing up in local community events. Actually, he had moved his family to the region to be close to the office and had developed a significant network through his children's schoolmates. All of the people he met could notice his expressed concern regarding any issues they had with the project, and he had widely distributed his contact details so concerned members of the local community could speak with him. He had already taken into account some concerns from local farmers and had changed a couple of truck routes to minimize disturbances in residential areas.

So he was not overly astonished when he received a call late that night on his mobile. It was from the mayor of one of the small villages nearby. As he hung up, his wife could not resist remarking on the fact that he looked worried. "There seems to be an issue that might impact our project," Simon added without further explanation, and then went into his study to make a few calls.

Early in the morning the very next day, Robert received a call while having breakfast at home. He was astonished to hear the caller present himself as the project manager for the tunnel construction project. Could he drop by for a face-to-face meeting to better understand Robert's concerns? Obviously Robert could not say no, and they agreed to meet that same afternoon.

Simon had now mobilized a large part of the network he had patiently built during the previous months, giving calls to many opinion leaders and politicians in the region. He showed his concern and at the same time committed to looking into the issue as speedily as possible so that he could better understand what was happening. The local press mentioned the issue, but it stayed low-key.

That morning, Simon got his team to look at the issue. One of the geologists happened to have a Masters in marine biology, which had been his passion before he decided to specialize in geology for career reasons. He let it be known and readily volunteered to help solve the problem. Simon was delighted to find such competencies in the team, and nominated him immediately to a leadership role in charge of the technical side of the issue.

In the afternoon, Simon listened, intently and with empathy, to the mussel growers who had met together in Robert's house. Despite being understandably emotional about the issue, they felt from this encounter that they could have confidence in Simon's ability to get things resolved. He had such a different approach to the problem compared to most technocrats they were used to seeing, who generally concluded coldly that the locals would get reimbursed by the government for any losses, anyway, and who then would brush the matter aside as though it were unimportant. No, they felt Simon definitely understood what it meant to them, their livelihood, their expectations of life, to be able to grow these mussels with love.

Simon decided to slow down work at the dig site, and even stopped the workers for an entire day, to ensure that all the environmental protections they had planned for were in place and remained effective. He had additional measurements taken in the surrounding land and in the nearby sea, looking for traces of the site's activity. And he allocated a survey boat the project had rented to help the mussel growers save what they could from their breeding grounds.

In the meantime, on the other side of the strait, Alan, the Norlander's Mussels Growing Association president, had received a worried call from Robert. He mobilized his colleagues immediately to go to their breeding grounds and to check the health of their mussels. And indeed, they soon found that at some sites the mussels had suffered a sudden and unexpected death. As Robert had done before him, he brought the issue to his mayor and representative, and spoke about it to any journalist that would listen.

One week later, Hannah ran into Harry's office with a worried look on her face. "Look at what I found," she said, unfolding a newspaper. In thick black letters was the main title spread: "Sea Strait Tunnel Kills Mussels by Thousands," with a few gory pictures of expired mussels below for maximum visceral effect.

Hilary came running in a few minutes later with bad news: the work site had been blocked by a demonstration! Work could not proceed that morning!

"What the hell?" Harry demanded, completely surprised by this unexpected event. "What's happening here?"

The phone rang. "Harry, this is Ben. What is this mess all about? The Prime Minister is concerned that the project is being challenged. The Prime Minister will have to answer questions about this issue in the House of Representatives later this week. Harry, how come I was not informed that there was a serious issue regarding the dig site?"

Harry was as puzzled by this sudden event as Ben was, but he had to maintain the impression that he was in control of the situation. So, he answered, "Yes, that's a problem we are working hard to understand. I mean, we've done everything prescribed by the environmental agencies...."

Ben cut him off. "Harry," he said, "get your ass out of your nice office and go sort this thing out before it becomes even more politically sensitive." Then Ben hung up. Harry looked at the abruptly silent phone handset in disbelief. That was clearly an order!

During a brief, puzzled silence Harry looked at Hilary and Hannah. Then he started screaming instructions, trying to get the team moving to solve the issue.

Once at the site, Harry found the issue difficult to solve. The crowd was very excitable and seemed opposed to all of the reasonable solutions he could think of. First, he declared that the site could not have anything to do with this issue, which only further incited the crowd's fury. Seeing that it was not getting him anywhere (and after another furious phone call from Ben), he proposed a second reasonable solution: the project would compensate the local mussel growers for all their lost production. The mussel growers were upset about this proposal. Before leaving the demonstration, they shouted: "We won't let ourselves be bought out by the Government!"

The issue was finally resolved a few days later, when Simon's team managed to find out that the sudden death of the mussels had been due to the invasion of a particularly invasive variety of algae that would grow inside the mussel's shell, weakening and then eventually suffocating them. As the finding was corroborated a few days later by a highly regarded scientist who had been contacted for the occasion by Albert, the Oldavian mussel growers got on with their lives. Luckily it was possible to treat the issue, eliminate the spread of the algae to young mussels, and thereby avoid further destruction of the mussel growth areas. Robert became one of the warmest supporters of the tunnel project. He even brought freshly prepared mussels to the next project team evening dinner!

As the news spread to the other side of the strait, there was some confusion. Alan called Robert, wanting to know whether the news was real. Once reassured, the mussel growers lifted their occupation of the construction site's entrance. But they would remain forever suspicious and were no longer supportive of the project.

When it was all over, Harry was baffled to learn that the mussels problem, as he would later refer to this issue, had started in Oldavia, but had not affected the Oldavian project very much. He was even more surprised when he learned that the solution to the mussels problem had also come from the Oldavian side of the strait! He felt that the solution had been completely out of his control, and he felt overwhelmed by the situation in general. For Hilary, the analyst, what had just happened was even more mysterious; she could not decide how to implement such an event in her highly elaborate project models. It was almost as if Martians had suddenly disembarked on her planet; the data simply didn't fit anywhere.

Simon had followed the controversy on the Norlanders' side. As he held a town hall meeting to debrief everybody on the outcome, people in the team were wondering why there had been such a difference in the depth of the crisis in the two countries. After some thought, Simon said, "We've heavily invested in weaving a solid network of stakeholders. It was an investment which has been well repaid in the last month. Thanks to this network, we have benefited from an early warning and from all the support and sympathy that we needed to resolve this issue."

The next day, somebody had hung up Simon's networking principle on the project team's office wall:

Invest in weaving your network to overcome the inescapable difficult moments

---

-14- Progress measurement hiccups

As the worksites' construction was now progressing at full steam, Hannah and Sally on each side of the strait were running their project controls processes, updating their progress in the schedule and developing forecasts that would be used by their respective project leaderships for making decisions.

In the Norlanders' office, the work was closely followed and coordinated by Hilary. She was running the most impressive task assignment program. The central schedule would issue the workers with to notifications of when to start work, and notifications of when they were late. The sheer precision of the Central Schedule led to the issuance of dozens of such notifications every day. The office was efficient and largely silent, with everybody concentrating on producing the deliverables as per the overall plan. Progress and performance indicators were regularly updated on posting boards. The team members felt the pressure and were working hard to deliver, each of them absorbed in their particular task, focused on their computer, and isolated by their earphones.

Harry was essentially invisible to most of his team members. He spent the greatest portion of his time in following the dozens of analytics reports delivered by the project management system to a large-screen dashboard in his office. At the same time, he did not discuss these numbers regularly with his team. He instead spent a lot of time explaining the project to journalists and appearing on many television interviews. His ego was stroked. He could not believe the amount of attention he was getting, after more than thirty-five years of building tunnels in the darkness of anonymity!

Hugh, the project management consultant, was excited about the availability of all the data for his research, and found himself nearly trembling whenever he thought about how perfectly set up this project was. It had the potential to become a source for decades of case study analyses, enough to supply him with material for the remainder of his academic career.

Hannah's project controls systems were extremely detailed, and they quickly proved to be very painful to update regularly. As Harry wanted to have his dashboards updated at least once a week, and sometimes wanted updates even more frequently in the case of particular tasks, this soon caused trouble. The project team had to hire several planners and cost controllers for the sole purpose of updating the reports of progress on the thousands of activities and budget elements that had been identified at the beginning of the project.

Hannah's worst issues had to do with changes to the initial plan. They impacted so many different elements and affected so many systems that implementing each change in a consistent manner was very nearly a project in and of itself.

Data gathering and coding became a nightmare, and errors and inaccuracies started to appear. And what inevitably would happen, did happen: updates were made later and later, errors developed and were perpetuated, and the references on which Harry based his decisions and Hilary based her analyses became increasingly flawed.

As they grew more conscious of the issue, Hannah and Hilary had had a few discussions on simplifying the project tracking systems. When they mentioned this to Hugh, however, Hugh defended the level of detail required by the system and explained that the inaccuracies encountered might simply be the result of a lack of precision in both the breakdown and the follow-up. He instead pushed for adding even more categories and activities to the initial plan!

As confident as Harry felt about his own technical skills, he was in awe of Hugh, who was after all a recognized academic professor. So Harry agreed with Hugh's assessment of the situation, Hannah and Hilary could only comply with Hugh's requests after a vain attempt at further resistance. As a result, the follow-up system became more and more detailed—and increasingly unmanageable.

The adverse effects brought about by the excessive details required for input into the system were not obvious at first. The main activities that everyone focused on were closely updated. But due to the lack of time, secondary activities did not receive the same amount of attention, and errors could remain in the system for a while before anyone noticed a discrepancy.

At one point, a set of obscure welded components called 'locking discs' was required in order to lock the steel supports of the shaft walls in place. They were quite simple components to manufacture, and a small supplier had won the contract to produce them on a low bid. The supplier was sending reports of production and quality control to the project team on a regular basis, which looked good as far as the team's members were concerned; nobody was really interested in paying attention to reports involving such a menial component.

As the shaft was being dug, Harold, the worksite manager, asked about the locking discs, as they would be needed in the next few days. These discs were due to have been delivered a while ago, according to the latest production schedule. Harold investigated, and discovered that they had not been physically delivered on site. There was some confusion, and an argument arose between the buyer and the storehouse; the buyer would not believe that the locking discs were missing, as it was not what the system had said. Harold nearly had to ask the buyer to come and find them himself before he could convince him that the discs were simply not there!

If these components were not made available for use within one week, the shaft digging activities would have to stop, as the shaft could not be safely dug much deeper without the additional security provided by these disks.

The issue was promptly mentioned to Hannah, the project controls manager. Hannah did not understand the issue; there seemed to be no problem in the schedule. The other systems gave conflicting indications as to the status of that particular order. Hannah did not want to bother Harry about this, as the component in question had nowhere been identified as a risk, and so nothing got done for another few days. Anyway, as a perfectly sensible person, would you want to talk to Harry about such minor problems, in particular if it would create the appearance that you were not adequately completing tasks in your area of expertise? You would be executed on the spot! And after all, with a little bit of prayer, problems do tend to solve themselves, don't they? Following this general philosophy, which was now an unwritten rule which most of the team followed, the issue got buried.

Unfortunately, the issue only became more pressing instead of disappearing into the aether. After a few days, Harold became courageous enough to call Harry directly, and informed him that there was now a strong possibility that digging the shaft would have to stop within three days unless locking disks could be supplied! This finally created some sense of urgency in the office as Harry's temper exploded on the spot, and he loudly uttered a long list of swear words that were heard throughout the entire project office.

After several attempts, someone managed to get hold of the supplier's representative, who spoke to them from next to his swimming pool as the plant had closed for annual maintenance and repair. The supplier had not finished the work, but was not overly concerned about the matter. Given that the locking discs were a low margin product, the matter had not received a lot of attention. Nobody had told him what the components were for, or whether they were really important.

Now that the pressure was increasing, a team from the project office was sent to the supplier's facility, where the components were discovered in a dusty corner of the workshop. They were of poor workmanship, and clearly below what could be considered an acceptable level of quality for such a critical component.

The crisis escalated. Work stopped in the shaft. Expensive welders were mobilized from the best welding company in Norland to remedy the situation. The entire project team became idle, waiting every day for the progress report from this emergency response team on which all their hopes had been laid. The welders worked day and night, and accomplished a miracle. The first pieces were delivered one week later, and the shaft work could start once again.

Having left Harry to stew in his office over the incompetency of his team and how upsetting the world, Hilary and Hannah sat at a computer together to try to understand what had gone wrong in their extensive systems and models. How could such a comprehensive system have failed? The systems were actually not to blame; they just had not been updated properly. Because people had not had enough time to update their reports with all the details required by the various systems, they tended to anticipate and extrapolate past progress to keep Harry happy. As Harry did not meet regularly with his team to discuss their progress and upcoming events, he was satisfied with nice looking numbers that were never really challenged—which was fine for most of the team!

Harry would not listen to their analysis or their suggestions for simplifying the system as the project moved forward. According to Hugh, what had been implemented was sufficient for the project's needs, so Harry just concluded from these recent events that Hannah must be utterly incompetent, and therefore had to be let go. He brutally told her not to turn up at work anymore, and ending their meeting without allowing further discussion.

After an epic struggle, Hilary, who did not see herself taking over Hannah's previous responsibilities, managed to bring Harry around. Hannah was allowed back in the office. The project continued on, but Harry was now mistrustful of Hannah's reports and systems updates. Hannah's mood was sour, and she no longer felt motivated to keep her performance above the minimum acceptable quality. Soon, the team became even more shortsighted with regard to the actual, real-world status of the project.

On the Oldavian side of the strait, Sally had chosen from the very beginning to limit the level of detail that was included in the planning and controls systems, preferring to be sure to keep a grasp on what was most important. In addition, she was aware that the main risk in a large, complex project was a convergence failure, where secondary components or deliverables are not there to meet with the other required deliverables at a major convergence point, and so the project grinds to a halt. Hence she implemented a convergence planning process that monitored the project's main convergence points in advance. The required deliverables were listed, and the available float was monitored regularly. They had had to take action with some secondary suppliers to avoid getting caught like the Norlanders had, and they had been successful in avoiding any delays on secondary supplies that could have stopped the work that was being done at the site.

Simon was very disciplined about participating in regular reviews of progress, costs, and risks. Even when travelling, he would make a point to always study the documents in advance and participate actively to the meetings by teleconference, challenging numbers and questioning trends, digging for the underlying causes behind the results. He was so disciplined, in fact, that even Sally was surprised by his response one day when she suggested postponing a monthly review. She did not expect Simon to be so unyielding with regard to this issue when he was so adaptable with regard to most others.

Simon had to explain to her that it was important to maintain regularity and discipline in the underlying controls of the project, and he wanted to show everybody how important this matter was to him. He explained, "Sally, these are my instruments which allow me to understand the current situation of the project I'm responsible for. I have to look at them regularly and recalibrate them if it becomes necessary—I can't afford not to. I don't want to find myself surprised by unexpected events that we should have seen coming. So we need to do it, regularly, just like clockwork, and I will not accept any compromise on the updates of the few indicators we've chosen as critical."

And the monthly reviews became part of the DNA of the team, with regular reporting bringing together the relevant information, as steadily and as dependably as clockwork.

-15- Mutiny looms on site

Work at the dig sites was hard. Teams were working around the clock, and the weather conditions at this time of the year were often poor. When it rained, the water seemed to come down in buckets, and afterward the sites looked like little more than large puddles of mud.

Simon had followed a specific strategy for choosing the crew of his construction team. He had taken a core team who had already worked together on large infrastructures in his previous projects, and had complemented this core team with tunnel experts.

Sandra, the surveyor, had gradually risen to be the young star of the team. In the past months, Simon had been particularly impressed with her. The young surveyor was showing impressive leadership skills. It had taken a lot of effort on Simon's part to push her to take the challenge of stepping forward and leading progressively larger teams, but it had paid off. Although she was a young woman, she could get the old gray-haired front-line workers to move at an incredibly consistent pace while maintaining a fantastic atmosphere. They all were in love with her. Seeing this, Simon had promoted Sandra quickly. At the beginning, Sandra had made a few missteps, and Simon had asked Scott to provide one-on-one coaching to help her. He also made himself available for mentoring. Her progress was incredible.

Sandra was always careful to watch over the well-being of the team. Their working conditions were hard, but she had set up inviting-looking coffee kiosks so that the workers could have warm beverages, sandwiches and pastries when it was cold, and overall she had taken quite a few steps to try to make the workplace more bearable.

Because the project office was intentionally very close to the site, Simon was able to organize several "project parties" involving both the site workers and the office staff. These parties were specifically designed to get the people from both sides to know each other by doing challenging activities together. Simon's philosophy was very clear: by getting the people to have fun together, they would achieve much more as a team when difficulties inevitably arose.

Moreover, Simon was adamant that his team's engineers would go daily to the dig site's work areas to discuss issues with the workers. A tracking table had been set up, and the engineers' on-site visits were one of the main indicators that the project leadership followed. Simon also invited Sandra to all of the project core meetings, and Sandra would bring along whomever she thought could contribute most to the particular problem at hand.

The Oldavians' construction team suffered through the rainy weather, found ways to celebrate, and dug their way through the mud at a steady rate.

On the other side of the strait, the Norlanders had brought together a team of experienced construction workers who had a great deal of experience in constructing tunnels. They had joined the team one by one from all across the world, and they generally did not know each other from previous projects. Harold, the site's supervisor, was a giant of a man. He was tall, wide, thick, muscular, and generally imposing; he claimed to be descended from the Vikings, and few people cared to challenge that claim. When he shook hands for the first time, one could hardly help but wonder whether his large, muscular hands could not crush one's own. Not many people had the courage to debate the instructions he gave, as they were always delivered in a strong voice, and in a tone which did not invite discussion. At the same time, he was always ready for a good joke and a beer—when everything was going well, at least. And he was always ready for a shout when things were going south.

Harry and Harold did not get along very well together. Harold felt that Harry was a kind of remote dictator with a team of intellectuals who were unable to understand the issues of a construction site. Harry, on the other hand, thought that Harold was just a dumb, uneducated brute who was only good for following the detailed instructions produced by the project management program. Neither one of them made much effort to meet or to discuss key issues with one another.

Harry was pushing Harold on progress, and Harold, in turn, was pushing his crew. Still, Harold knew he could only push his team so far. The working conditions were often rainy, miserable or muddy. The crew did not know each other very well, and did not communicate with each other readily. Each department and each section continuously blamed the others for making mistakes. Each small problem required Harold to intervene personally in order to get people working together to find solutions. From the outside, progress looked good, but Harold knew that it came at the expense of wearing down the team physically and emotionally. Some of his supervisors had not been able to sleep for two days when there were issues that needed to be solved. Harold himself was on the worksite nearly twenty hours a day.

The Norlanders were almost done with the digging of the shaft—they had only one week to go—when one of their main pieces of equipment broke down. It was annoying, but quite to be expected for those experienced hands who knew that they had been driving both themselves and the equipment hard during the past weeks. Harold considered the issue to be nothing out of the ordinary. A few days would be needed to get the parts and then get the machine fixed, but that was fine. It was a good opportunity for some of his supervisors and work crews to take a break and get some well-deserved rest.

But Harold hadn't counted on Harry. As soon as the news reached him, he was on the phone with Harold. He sounded upset and stressed, and this unexpected drawback was pushing his temper to the limit. Harold tried to explain how this was really a fairly normal event; equipment breakage and repair was part and parcel of the life of construction workers, and something that was generally planned for as part of the site's downtime.

Harold was astonished when Harry's temper erupted. "That is not part of my plan. What happened? You forgot to do maintenance, or what?" Oops! Harold looked at his phone in surprise, as if he was wondering whether what he just heard had actually been said. But Harry didn't stop there. "Don't stop working! Dig with your hands if you need to! But keep moving forward! We can't afford to be seen doing nothing!"

Harold tried to reason with Harry, but it was to no avail. Harry finally cut the conversation short by repeating his previous instructions and promising all sorts of awful consequences for Harold if he did not obey.

When Harold put down the phone, he was regretting for the first time that he had ever accepted this job. He called the supervisors in. Once they were all together in the room, he looked at their tired and worn-down faces, and with a sigh, conveyed Harry's instructions to them. After a short, astonished silence, the supervisors' voices rose, and a few of them started gesticulating to make themselves understood despite the noise. Their meaning was all too clear. They were fed up with working on this dumb project that afforded little sleep and high stress, and which gave them no recognition whatsoever. Who was this guy, Harry, after all? They had never seen him in dirty boots and coveralls, never seen him arrive at the work front to chat with them like teammates and equals. How could he give them instructions like that?

A fierce debate ensued, and by the end of it, a majority of Harold's supervisors had decided to stop working in protest. They were fed up.

Harold looked at the scene in front of him in disbelief. He had mixed feelings about the matter. He could not disagree with the team, but he also felt that he had to be seen as the boss. When the supervisors had finished expressing their position, they turned to him. Briefly, total silence fell over the room.

Harold knew he had to take a position. And he had to do it now. He looked at the supervisors, one by one, and spoke.

A few minutes later, Harry got a call. He had not yet hung up before he was already screaming, and the whole office could hear him. "A mutiny!" he roared. "We practically pay these guys in solid gold, and they still refuse to work! I am going to fire them all and get new ones!" he screamed. Hilary and Hannah stared at him, surprised. They had been trying to analyze the delays caused by the breakdown in equipment. After a while, Harry finally became calm enough to explain to them what was happening. That is, he gave them half of the story—his half.

Soon the word spread that there was some hiccup on the Norlanders' site, and Harry could not avoid going there to manage the situation. He bravely brought Hilary and Hannah along with him, as reinforcements.

The discussions at the site turned out to be quite difficult for Harry. He and Hilary had to endure a verbal bashing from the site team for having developed a plan that was not realistic (of course, they had not been consulted). At the same time, Hannah figured out that there were a number of people and pieces of equipment at the site which had not been accounted for in her project control system. She winced as she realized—because of this discrepancy, there would be an unexpected cost degradation in the next reporting! And that prospect made her deeply uneasy.

Finally, after long hours of talks and negotiation, Harry agreed to review the plan and to give the site supervisors some money to increase the workers' level of comfort during miserable weather. Work started again a few days later when the shaft-digging machine was fully repaired.

A lot of people at the site started talking amongst themselves about how they were pressured, how they had to take shortcuts, and how these shortcuts had resulted in a few accidents, all because of the unrelenting pressure from Harry's team in the office. For the first time, the journalists and newspaper editors were not sympathetic to the Norlanders. Headlines like "Tunnel Labor Exploited" and "Tunnel of Blood" quickly got the attention of the public.

Hearing about these stories, Simon discussed the issue with Sandra, who was now the site supervisor, to be sure that the same thing would not happen on the Oldavians' site. Sandra was very much at ease and felt confident that those issues would not happen. She knew the energy level of the team was fantastic, thanks to the efforts of everybody to develop a strong sense of team spirit. Indeed, because it was a small world, quite a few of the Norlanders' hired hands were now candidates for the positions available on the Oldavians' side, and many of them looked extremely promising. She concluded by telling Simon, "Here's another thing you can put on the wall of this office. I learned it from you." She smiled, and then said, as if quoting, "The way to do extraordinary things is not to bring together extraordinary people. It is to bring together ordinary people, and leverage the team to do extraordinary things."

The next day, somebody had posted that sentence in large letters outside Simon's office.

> **The way to achieve extraordinary objectives is not to bring together extraordinary people. It is to bring together ordinary people and leverage the team to do extraordinary things.**

When he heard the news, Albert could not resist calling Ben. "Hi Ben," he said when his friend answered, "it looks like you have had some labor relations issues with your crew. Do you want some help? Generally in Oldavia we are quite accustomed to having organizations or teams strike, and we have developed a certain flair for continuing to get work done nevertheless. If you want, I can send you one of our experts...."

Ben declined politely. As he set down the phone in its cradle, he had to wonder. How could these generally disorganized and lazy Oldavians be so self-assured?

-16- A grain of sand in the plan

Although the Norlanders were delayed by issues involving the work crews and the damaged equipment, they got ready to bore the tunnel more quickly than the Oldavians, and so the Norlanders were able to start boring the tunnel earlier. This was great news for Harry, who phoned Ben to bring him the good news.

That was when disaster struck. In the middle of this well-oiled machinery, of these rapidly rotating wheels, a grain of sand caused an enormous mess. A grain of sand? Well, actually, a large, bulky boulder!

"What do you mean, 'We found a boulder and our tunnel-boring machine broke down?'" thundered Harry, walking to and fro in his office, casting angry glances at the engineer who had brought the news. The engineer looked miserable. His whole demeanor suggested that he expected Harry to bring an axe out from behind his desk and to have the engineer's head chopped off in the next minute. After a while, he found the courage to explain that finding a boulder was unexpected in this particular geological layer, and that the machine had not been designed for removing boulders. The boring mechanism just could not handle it, and was stopping repeatedly. The engineer concluded, "We can't continue like this. We need to find a different solution if we're going to dig this tunnel."

Hilary looked up from her computer. The dashboard of the project, which she followed continuously, was already showing that progress had stalled. Indeed, looking outside her cubicle, she could see that most of the team was standing at their desks, glancing over their partitions, looking anxiously toward Harry and the engineer.

Harry realized it. He stepped outside the meeting room and yelled, "What are you looking at? Go back to your work!" Silently, all the team members retreated safely back behind their partitions.

Returning to the meeting room, Harry looked at Hilary. He suddenly looked much older and more tired. "So, genius, what's the plan now?" Harry asked.

Hilary stuttered: "I don't know... err... We need to analyze the issue and totally rethink the project, I suppose! That's something I did not take into account," she added, typing fiercely on her keyboard.

Harry looked overwhelmed. In a weak voice, he told Hilary, "Just bring me the solution when you have it." He went back to his office, looking as if he didn't have any more energy left with which to fight.

Hilary was dumbfounded by Harry's sudden change. The unknown cannot be foreseen, after all. And the Oldavians were digging in the same layer. They would find the same problem. And they were not even half as efficient as the Norlanders.

The Norlanders would re-plan the project, Hilary knew. They would do it efficiently. And they would win. It took only a few hours for Hilary to set up a plan. As Harry was absent from the office, she brought the team together and explained the new plan to them. "Get on with it," she said as she finished. "Don't sleep until you get this freaking plan up and running".

The team was dubious. Digging though the boulders would take time. Sure, the technical people had a solution, and that was better than nothing, but it would be really painful to implement. They went back to work half-heartedly.

-17- Removing the grain of sand from one's shoe is better than walking with it!

Because of the network that Simon had woven and which he continued to expand and reinforce, the news of the Norlanders' mishap with the boulders reached the Oldavians only minutes after the event happened, and the information came directly from the front line.

When the news reached Simon, he immediately convened a meeting of his core team. Sandra was now part of it, of course. He brought up the issue. Steven, the tunnel expert, looked extremely annoyed to hear about the boulders.

"If this is accurate, we have a major problem on our hands. Digging the tunnel with our current equipment under these conditions will take ages." He shook his head, then he quickly added, "I remember having been in this situation about twenty years ago. The construction just dragged on and on. We almost had to dig the entire tunnel by hand, with small spoons. Luckily we just encountered a short stretch like this."

The team went immediately into solution-finding mode. Sally stood up and went to the board. "Let's brainstorm," she said. And within a minute, they had all stood up facing the board, and were sending ideas out while Sally wrote them down as fast as they arrived. The core team called for the most recognized experts of the team to participate, too. Sandra got the most experienced people from the dig site to join, which they did a few minutes later, leaving their muddy boots at the entrance of the office.

Simon felt proud of the team—although he almost felt he was of no use among such experts! Deep inside, he knew that all the work he had done to get the team to work together as one team, making sure each member was emotionally engaged in the project, and investing his time in helping his team develop a positive attitude, was now bearing fruit.

The most strange and unconventional ideas were thrown around. Spontaneously, the entire team had joined together to solve the problem before it became a major disruption.

Sally facilitated a number of rounds of brainstorming, which were followed by categorizing the options and then several rounds of voting. By the end, the many suggestions had been whittled down to a few main ideas.

After a few hours, the result emerged. It was simple in hindsight, but hadn't been easy to find for this team of makers and doers who were generally used to dealing directly with whatever condition they found. But rather than forcing their equipment to struggle with digging through the boulders, the best solution was to dig around them—or rather, to dig in a slightly deeper layer, should the geological tests confirm that such a plan was appropriate. The idea had actually originated with a graduate engineer as part of a brainstorming round. Although it had initially been dismissed, it had somehow survived several rounds while ideas converged and the team members debated. Eventually, it found more support, and appealed to a large percent of the core team. Once the core team members had decided on a course of action, they breathed sighs of relief and let themselves laugh a little at how concerned they had been. Their solution seemed so simple in hindsight!

Some tests were carried out and checks were quickly made, with adjustments made on-site in the shaft. A full-fledged management-of-change process was carried out to ensure that no aspect was being overlooked that could derail their new plan. After all of their specialists gave them the green light, tunnel boring started in the new layer. The team's energy was high. Now they were certain—the Oldavians could win!

-18- Crushed dreams

On the Norlanders' site, the solution that Hilary found was to make extensive modifications to the tunnel-boring machine so that it could deal with boulders. Now, that plan required disassembling most of it and bringing those parts back up the shaft. Unfortunately, the shaft proved to be too narrow for them to implement the quickest solution, which was to dismantle as little as possible of the machine.

The Norlanders thus lost precious weeks designing and implementing the modifications to their machine, and then lost further days in bringing parts up and down the shaft. In the meantime, the Oldavians had finished their shaft, installed their tunnel-boring machine, and had started boring.

Harry was confident that the Oldavians would have to deal with the same nasty surprise that had delayed his team, so he kept repeating to Ben how clever he and his team were, and added that they had found the best technical solution. It would take time, but they would easily overcome nature's unexpected hurdle. Harry was enjoying himself, pleased about the surprise the Oldavians would find, as he had made sure that the discovery of the boulders had not leaked to the press.

As the news began to report that the Oldavians had started boring their side of the tunnel and were making a great deal of progress in a short time, Harry could not believe what he was reading, much less understand how they had done it. They surely should have encountered the same natural obstacles; he had been led to believe that the boulders would be found at both teams' dig sites! Nevertheless, progress was now being made on the Norlanders' side, as their equipment dealt with one boulder at the time. Every day, every hour, Harry urged his work crews to go faster, faster!

Team spirit was high on the Oldavians' side of the strait. They had already covered a quarter of the overall tunnel length when the Norlanders had done only one tenth!

Without warning, the main gearbox drive for the tunneling machine broke hardly a day later, failing with a deafening sound of crushed wheels and mangled hopes. When the news reached the surface, Sandra immediately called Simon. He could tell that she was almost in tears! As an experienced construction project manager, Simon had sensed that things had gone a bit too smoothly lately, and had expected some disruption. He hadn't expected such a serious issue as this, through!

A crisis committee was immediately brought together to deal with the problem. The team's enthusiasm had melted away in the wake of the disaster; it quickly became obvious that this was a potentially game-losing situation. Although spare parts had been prudently ordered ahead of time and were available to be installed, the boring machine was now ten kilometers away from the entrance shaft, and there was no other option than to bring back a large part of the machine to the dig site. There, the mechanics would be able to do the necessary repairs.

The faces around the table looked tired and tense. Several options were discussed and summarily dismissed. There wasn't much choice as to what needed to be done.

After a quick discussion amongst the old timers, led by Steven, Simon's experienced deputy, Steven announced sternly that the repairs would take at least two months. "And that's if everything goes well!" he added.

The team's spirits sank further. Two months! Meanwhile, Sally had been plotting the Norlanders' slow-but-steady progress. She drew a few lines on Simon's chart, looked up, and announced in a steady voice, "With two months spent making repairs, even if we progress as quickly afterward as we did before our equipment failed, we won't make it to the middle of the tunnel before the Norlanders do."

A long silence ensued. Steven was the first one to break it. "Sally, how quick would we need to be for the repair to still have a chance to win?"

Sally looked at her charts, drew a few more lines. "I would say we have roughly one month to make repairs and still have a chance—*if* the Norlanders don't accelerate, and *if* we manage to continue boring at the same speed afterwards."

Everybody looked at each other. "One month! That's impossible," said one of the more experienced people in a low voice, verbalizing what everybody was thinking. "That has never been done."

Soon afterward, in Norland, Harry was all smiles. The news of the Oldavians' disaster had just reached him. For once, he looked relaxed and pleasant. "That's it, they are done!" he announced confidently to the rest of the office, passing his index finger across his throat to underline his message. "We're bound to win now. It's impossible for them to recover from that one!"

Hugh, the project consultant, was beaming. He had feared for a moment that his scientific methods of project management could have been challenged by the horde of disorganized barbarians on the other side of the strait. Only Harold, the site manager, who happened to be visiting the project office that day, was shaking his head in negation. He knew that such a problem could also happen to them, and that if it did, his crew did not have spare parts; they hadn't been ordered for cost-cutting reasons. He felt sorry for the Oldavians.

Simon was taken aback by the incident. In an instant, his high hopes had tumbled off a precipice. Nevertheless, he was the project leader, and his job was to demonstrate energy and enthusiasm, which he went on to do, no matter how costly that would prove to be. For a while he had to hold the team together with his high spirit. "Would Oldavians lose without fighting until the end?" he asked his team. "Would we let such a menial challenge stop us from succeeding?" These were some of the messages that he passed along, and soon, his optimism had brought energy back into the system. Slowly, morale improved.

The team went down the shaft into the tunnel. The silence was eerie in a place where the noise of sold crunching and moving would normally be deafening. The now-opened gearbox looked and smelled like a stack of half-molten, vaguely round metal shapes. It was quite a mess! The moist darkness of the tunnel gave the scene something of an end-of-the-world atmosphere as everyone was plodding their way through mud and puddles. Simon paused to give comfort to all the workers he met as he passed through. "We will win; I'm confident that you'll make it work in no time!" was the message he delivered consistently to every person he met.

Albert had been, of course, continuously updated as to the current status of the situation. He felt very poorly. It seemed the gods were not with him, or with his moustache! He admired how Simon had always stayed very professional before and after the disaster happened, staying factual and prudent. Well, obviously Simon did not have a precious moustache to lose! However, even his decades of experience in diplomacy did not prepare Albert for the call he received from Simon four days after the breakdown happened.

"Albert, we think we might have a way," Simon began, "but we'll need your help with something."

Albert was instantly all ears. He had just sat in on a meeting of a committee of expert tunnelers, and all of the most renowned experts he had brought in for independent advice had shaken their head or looked down, leaving Albert but little hope of saving his moustache. What were these guys up to now?

"You see," Simon explained, "thanks to the design of our shaft, we can manage to bring a large part of the tunneling machine back up instead of doing the complex mechanical work in the tunnel, which would mean working ten kilometers from the entrance and dealing with grueling conditions. The guys have been brainstorming various scenarios in the last few days, and we played with the model, you know, the one in the middle of the office. And because we have a lot of spare parts, we can actually prepare a lot of things in parallel. It might or might not work in time, but I believe it's worth trying. My guys are all excited."

Of course that was worth trying! Albert thought. Excited, Albert immediately answered, "Tell me what you need. This is a national priority. If there is even a small chance that we won't let the Norlanders win this time, we have to take advantage of it. We'll do whatever it takes!"

During that time, the Norlanders were grinding their boulders away one by one, and slowly but surely making progress. Never had Harry been so happy and so congenial with his team.

-19- Waking up the Beast

"I knew these Oldavians were mad up to the point of being ridiculous, but here they have exceeded the limits of what can be borne!" exploded Harry, wiping a tear of joy away from his eye. On the table, the paper had announced in bold letters: "Oldavians mobilize Army Engineer Corps's vehicles." Below the article, the paper showed an armored caterpillar vehicle being lowered into the shaft. "Do they really believe this will help them? Do they want to shoot us when we meet them?" he asked mockingly.

He pushed the paper towards Hugh and Hilary, who looked at the paper's headline incredulously. Hilary observed, "Well, these guys have always surprised us, so I would not be so confident if I were you." Harry shrugged his shoulders and left the meeting room, still laughing.

Hugh turned to Hilary and said, frowning, "I think you're right; they might still surprise us. How do they come up with all these ideas? I don't understand it!"

It was a plan forged by a group, led by Steven, which had come up with this idea after extensive discussion, several rounds of brainstorming, and a lot of testing with the physical model. What they needed was a way to get a large chunk of the tunneling machine out of the tunnel as quickly as possible, but the normal tunneling equipment was only designed for crawling slowly, following along in the wake of the boring machine. They needed a pair of powerful, rugged vehicles that could tow the equipment out more quickly, and then bring it back in for installation once repairs were finished. So, they had come up with the idea of borrowing vehicles from the Army, a request which Albert was only too happy to help arrange.

And the Oldavians' site was organized like an army preparing for battle. Mechanics took turns, working in shifts all around the clock. Without the need to order anything, everybody had come back from leave, and a crisis organization had been set in place around a crisis control center in the project office. It looked like a center of command you might find in a space center, and was manned 24/7, as its members kept busy coordinating multiple activities in parallel. It was tough; bunks had been organized so that people could take short naps on the spot, without having to go back home. But nobody wanted to be anywhere else, as it was so much fun to be there rubbing shoulders with the others in the team. Everyone knew they were working to get the impossible done.

Even some of the supervisors from the Norlanders' team had been called in by their Oldavian mates and had defected over. Harry shrugged his shoulders when he noticed. "Let them go!" he said. "If they think they can win by taking our people, fine; we're almost done, anyway!"

Harold could only look at Harry in amazement. His buddies on the other side seemed to be having so much fun, he wondered why he was staying himself.

Albert came to visit and could hardly believe the buzz he found. People were running around having the time of their life. Maybe they wouldn't win, but they would have done everything they could to help. And they would later recollect how they fought against the odds. And fight, they did! They made progress, bit by bit.. The damaged section of the boring machine was brought out on the eighteenth day; an entire new section had been prepared in parallel, and was lowered into the shaft immediately. It was back in place and coupled to the boring machine on the thirty-first day.

To encourage his team to have fun, Simon had suggested to everybody that they should act as if they were struggling to deal with the problem and pretend that they could not start boring again for months. Because the team had become so cohesive, nothing leaked to the press.

Forty days after the Oldavians suffered their equipment breakdown, Harry was sitting with Ben in his cozy office, high above the capital city of Norland. "You see," Harry said, "they think they can work miracles, but they can't. You can't go any faster or fix things any differently than common knowledge dictates."

Ben looked at Harry with a hint of worry in his expression. These Oldavians were sometimes so unpredictable, appearing to border on miraculous talent the minute after they had appeared swamped in utter mediocrity. This had been shown again and again throughout history—they would slide up to the point where everybody thought they were lost forever as a nation, and then suddenly rebound with ideas that were shaking the world to its roots. And when Ben had called Albert that morning, he had found Albert a little bit too relaxed for his taste. "You are sure they can't catch up?" Ben asked.

Harry tried to banish all doubt. "Yeah, pretty sure. It's physically impossible even if they could start boring now. And we've already covered 38% of the total distance, while they are stuck at 25%".

One month later, the Norlanders had reached the 48% mark, and Harry was busy preparing a victory party, when he received a call from Harold. "Harry, we have a problem. You need to come to the tunnel immediately!"

Harry wondered what the problem was. All of their equipment seemed to be operating just fine, as he could see from the video transmission on his computer. Nevertheless, Harold seemed very alarmed, and would not tell him more, so, as he had some time to spare, Harry decided to make his way to the site. "Okay, then. I'll drive there tomorrow morning."

Just after he arrived, Harry was ushered to be end of the tunnel. It was the first time that he had actually gone in there. Booted and wearing heavy coveralls, they travelled in a small, open vehicle of the kind often used in mines. The boring machine was working tirelessly, making a huge noise. Harold screamed in Harry's ear: "We've noticed something strange in our monitoring systems, and the last time we've stopped to deal with a boulder we certainly did hear something otherworldly."

Harry was starting to wonder if he was caught in a science fiction movie; if so, the next thing they would tell him was that they had awakened an unknown monster that would engulf them all. He shook his head to clear the thought away, as Harold waved to the boring machine operator. The machine turbines gradually slowed down, and a few minutes later, finally fell silent.

"Listen!" said Harold. Harry listened. He heard drops of water falling from the ceiling, splashing on the ground, and in the distance he could hear the noise of vehicles in the tunnel.

"I don't hear anything strange," Harry said, after having tried and failed to hear anything out of the ordinary.

"Listen again!" said Harold, falling silent without further comment.

Then Harry noticed it: a deep, low rumble, coming from under the earth, sounding far away but gradually growing louder. "What's this?" whispered Harry, too afraid to understand.

Harold was not diplomatic about it, but said, almost flippantly, "Oh—my guess is that it's the Oldavians. They're close. Very close. Strangely, they seem to be a few meters below the level we're at. And they're grinding away like hell!"

Harry's face went white. "What? What do you mean?"

Harold looked at Harry with amusement in his eyes. "Well, they're probably like five or six hundred meters away, which means..." —he paused briefly for a quick mental calculation— "that they must be over the half-way mark right now. I hope they call us soon so that we can organize ourselves for when they meet us."

Harry's world seemed to disappear in an instant. He suddenly understood what had just happened, how the Oldavians had been able to repair their machine, how they had driven their tunnel a few meters deeper to avoid the boulders that had so plagued Harry's team. The knowledge stung even more when Harold told him that he had received a call from some of his tunneling buddies a few days before. They had spoken about the progress they had made, and armed with this knowledge, Harold had organized Harry's fatidic visit so that Harry would discover that he'd been beaten at the last minute.

The person who was worst hit, psychologically speaking, was probably Hugh, the project consultant. He realized that his scientific approach to project management was clearly not the best method of management. Being as he was a particularly scientific mind, when faced with compelling evidence that his theory had limits, he decided to launch himself into a new field of study, seeking to understand and analyze the other drivers of success.

Ben was furious when he found out. He had lost face. He had told the Prime Minister that there could be no doubt about their winning the competition, and he felt betrayed. Now he had to write a statement for the Prime Minister that congratulated the Oldavians! But Norlanders had a reputation for fair play which he felt obligated to defend, so he made the best he could of a bad situation, and publicly recognized the incredible feat the Oldavians had performed. The Oldavians, for their part, were kind enough to underscore how close the competition had been in the public statements they made to the press.

The Great Dinner happened a few weeks later. Albert went to a barber to get his moustache specially dressed up for the occasion. His moustache was practically beaming after a treatment with special crème. Albert invited Simon to join them. Facing Ben, Albert was diplomatic and delicate enough not to talk too much about the events of the past months. Instead, he brought the conversation around to the topic of "Soft Power."

"You see," Albert said, raising his glass of excellent and extremely expensive wine in preparation for a toast to the improved future relationship between the two countries, "when it comes to unprecedented challenges, soft power really makes the difference, and is often preferable to hard power. Our President wants to strengthen the ties between our countries. He has proposed an exchange of high-ranking civil servants. Why wouldn't you like to come to Oldavia for a few years to learn a bit more about soft power? And you could take the opportunity to teach us a bit about how you manage to be so efficient in some areas, too, of course."

Ben was hardly in a position to reject the offer. He could tell, now, that Albert was a real master of soft power. Yes, he told himself. He would learn, and spread the knowledge!

Closure

The work was finished. The final documentation had been produced. The sun had risen following the close of the last celebratory team party. That day was to be the last of Simon's days in the project office. Simon's heart was heavy with the thought of leaving that fine team of people. Many of them had already left, seeking other employment. Sally had made sure to organize a mailing list so that they could keep in touch and have many future old-timers' parties. Simon knew he could count on his core team to join him again any time he needed them, whenever he was ready to take on another challenging project. He had plenty of proposals awaiting his response, but first, he would take a break and spend some time with his family.

Simon was only half-surprised when he entered the office that day to the sound of applause. His entire team had arrived early to welcome him. He had had some inkling that they might be preparing something for his departure. Still, he was delighted to see that the office was packed. They were all there, both site and office people, and it even seemed that those that had already left had returned just so they could be there for that particular occasion.

Then Simon spotted an entirely unexpected face. Albert was also there! He came forward, stroking his moustache with an air of contentment. Sandra followed, carrying a farewell gift. "Simon, on behalf of everyone here, I would like to thank you for everything," Albert stated formally. "And not least for saving the moustache I took a decade to grow!" After a while the roars of laughter died out, and he added, his voice serious, "You know, every time I came here, I was intrigued by the maxims that were hanging on the wall. When I discussed them with your team last month, everybody said it was the best project leadership wisdom they had ever seen. And yet," he added, a twinkle in his eyes, "Sandra couldn't help saying that something was missing. And she was supported in this by many others." Sandra's face reddened. "She told me what she thought, and I found that I agreed with her. So, we added one additional maxim. Sandra, do you mind telling us what it is?"

Sandra cleared her throat. "Boss, you have done something wonderful for many of us. I'm probably the most prominent example, but I know you have done the same for many others, in different ways. You catalyzed us to make the best of our talents. You encouraged us to try things that were not part of our job description or that went beyond what we were supposed to do. You didn't limit us by what our degrees said we knew how to do. You used people's musical skills, modeling skills, programming skills, and many other talents, hobbies and skills that are generally not taken into account in a working environment. So, to represent what you are doing, we have added a maxim along the lines of: Catalyze the use of people's talents, and develop those talents for the team's sake."

"And also," Albert continued after a brief pause, while the team was forming a pathway way for him and Simon to reach the wall at the other end of the office, "I have added my own little touch. I have added a title that comes straight from my background in diplomacy. And there," he added, pointing emphatically to the wall, opening his arms in a sweeping gesture, "is the condensed wisdom of this project team."

On the wall in large characters were inscribed the following words:

Simon's five principles of Project Soft Power

Invest in weaving your network to overcome the inescapable difficult moments.

Never forget the rules of Focus and Discipline: the Two Plus One Rule.

Invest now in preventive measures rather than allowing yourself to suffer later.

The way to achieve extraordinary objectives is not to bring together extraordinary people. It is to bring together ordinary people and leverage the team to do extraordinary things.

Catalyze the use of people's talents and develop them for the team's sake.

Simon opened the gift. Inside the package was a plaque entitled "Simon's five principles of Project Soft Power." He smiled. They had captured it all. Maybe, he mused, he should write a book on project leadership.

THE END.

The Project Soft Power framework: the 5 roles of Project Soft Power

Introduction

In this section, we will investigate the practice of Project Soft Power more closely. Our extensive research has identified five key roles that need to be practiced, and in the interest of making them easy to remember, we have associated them with specific characters:

- The SPIDER, weaving his network cobweb;
- The KUNG FU MASTER, master of focus and relentless discipline;
- The ENTREPRENEUR, investing in the long term and building the project toward a purpose;
- The TEAM COACH, unleashing the team's potential;
- The PEOPLE CATALYST, revealing each individual's talents.

The next five chapters will examine each of these roles in turn. Examples from the Fable (Section 1) will serve to illustrate each key role.

As you read about these different roles, you will find that some of them feel more familiar to you. Others will feel more remote. This is normal. The self-assessment test in Section 3 will give you a more objective view of your current situation with respect to these roles, and Section 4 will discuss how you can improve your Project Soft Power effectiveness.

Accomplished project leaders always find a way to ensure that they play each of these five roles consistently throughout the project's execution. Get acquainted with them. They will be the cornerstones of your future success.

First Role: The SPIDER, Weaving its Cobweb

Like the spider, the successful project leader patiently weaves his network, the cobweb. Like the spider, the general idea is to efficiently capture anything that comes close to the project and which could influence its outcome—stakeholders in particular—and to manage them appropriately.

Nobody knows everything. Having an effective network is also a way to ensure access to far more wisdom and knowledge than a single person could ever possess.

Having an effective, functional network is a fundamental tool for the project leader. We are not speaking here of the social professional network that is recommended for career advancement and other mundane tasks. We are discussing a specific network that has to be set up specifically for the execution of a particular project.

Why stakeholder management is a mandatory activity for the project leader

According to the Project Management Institute's definition, a project is a temporary endeavor undertaken to create a unique product, service, or result. In essence, a project produces a change that disturbs its environment, creating instability around the previously existing stable situation. As with any change anywhere, most people around the project will look at it suspiciously, and may even oppose it overtly. The most dramatic projects will create a veritable army of opponents, although this effect is not related to the size of the project, but rather to its ability to create a disturbance.

"Don't worry about people stealing an idea. If it's original, you will have to ram it down their throats." This famous quote comes from Howard Aiken (1900-1973), an American scientist who was very involved in the field of computing at its outset. Opposition can be active or simply passive: the capability of bureaucratic organizations to resist any hint of a change can be daunting.

Let's face it: the most challenging projects (and the most rewarding, if they are successful) will have to deal with a lot of opposition, and the project leader's job is to address this adequately in order to allow the project to proceed. Successful project leaders cannot escape this simple fact of life. It is the same for successful leaders in any kind of remarkable endeavor.

Of course, there might be active project sponsors and other supporters of the project. The life of the project leader is made easier when these supporters can be relied on. Alas, the power of negativity and the large number of potential disturbing stakeholders is always dramatic, and the project leader always has to get actively involved in stakeholder management.

The development of a specific network for the project is a key activity for the project leader. This network must have some particular properties, and once established, the network must be nurtured and used properly by the project leader. Let us examine all these stages.

The cobweb: a dedicated network with a purpose

The purpose of the project leader's network is to improve the odds and conditions for success of the project, and to maximize the value created.

To achieve this, the project leader's network must serve the following functions:

- Engage the most critical stakeholders in two-way communication (we are using here the terminology 'engage' and not just 'communicate' to stress that the communication must be two-way);
- Allow the project leader's communications to reach as many people as possible, whether they are involved from near or far with the project;
- Serve as an early warning system and amplification system for low-noise signals related to events that might ultimately derail or otherwise significantly impact the project;
- Serve as a way to access knowledge that would otherwise be difficult for the team to access;
- Provide the project leader with a realistic and candid view on the situation and status of the project—which often involves a close network with the informal organization and with other stakeholders that might have very distinct views on the project's progress and achievements.

Network development thus cannot be haphazard, or left to depend on random social encounters. The project leader must go through the exercise of identifying all relevant stakeholders, internally and externally, prioritizing them, and then reaching those that are the most important for the success of the project, in order to achieve the five objectives exposed above.

When doing this exercise, which aims to build the most effective network (i.e., the smallest possible network that covers as many types of stakeholders as possible), diversification is key:

- Diversify the network as much as possible in terms of people's origins, situations, social status, and level within the organization. It is not effective to engage two stakeholders who are already very closely linked; instead, it is preferable to engage stakeholders that would not necessarily have any link in normal life outside this particular project;

- Engage stakeholders that are known for their candid or contrarian views in order to be sure that you will have access to diverse viewpoints during the project;
- Create a system where interested parties can give you their permission to send information to them (for example, by requesting to receive newsletters or email updates);
- Aim at engaging 'mavens,' people with a particularly strong, already-developed network that can be leveraged easily, rather than isolated people with a weak or poorly maintained network.

In addition to this identification and prioritization exercise, when meeting with people on worksites or at engagement sessions with stakeholders, the project leader needs to use his judgment to identify other people that might contribute significantly to the network's effectiveness:

- People not initially identified as influential on the course of the project, but that are indeed 'hidden' influencers in their community or in the informal network of their organization;
- People that are a reliable source of direct, candid information on the situation in the field.

Still, it is not enough to identify the key people that constitute the most effective network for the project, prioritize them to decide on the level of engagement, then and establish connections. The project network is a living organism. Like the Japanese game Tamagotchi, it needs to be constantly fed and nurtured in order to ensure it will deliver the expected contribution to its master, the project leader, and help to bring the project to completion as a result.

Creating rapport

While engaging with stakeholders, two golden rules need to be followed:

- the rule of giving;
- the rule of listening.

The 'rule of giving' states that creating a relationship always involves giving first—and, often, giving without any assurance of receiving anything back. The project leader needs to give in order to reach out to the people he intends to associate with in his project network. What should the project leader give? Direct access to him can be one key gift, and access to specific information of interest is another. The wise project leader who has already built a substantial network can also give something else which is often forgotten or overlooked: access to the other people in his network. This can be extremely valuable, for example in the case of reaching out to a grassroots opponent of the project, by giving him access to the institutional representatives that were already brought together in the project network.

The 'rule of listening' states that one of the best ways to create a relationship is not to communicate proactively, but rather to listen. Taking the time to listen deeply to stakeholders instantly creates both rapport and sympathy. It is not easy for the project leader to listen effectively, to give each stakeholder the amount of undivided attention that they deserve. Still, that's a skill perfected by politicians (and it is probably the only politician's skill that should inspire project leaders!), and it is a skill that can be practiced. Genuine listening, giving one's full attention to the person's issue, even for a short time, can go a long way in creating a relationship even if that does not mean an agreement has been reached.

One of the main goals of establishing a solid network, including rapport with stakeholders, is to allow negative feedback, criticism, candid situation assessment and other disturbing information to reach the project leader. This is fundamental as a way to establish an effective early warning system and to permit a candid evaluation of the situation as circumstances change. The successful project leader understands that maintaining an open channel of communication for these disturbing pieces of information is fundamental, so as to avoid being cut off from reality. To achieve this, it is vital to build an appropriate level of mutual rapport and respect with established critical stakeholders, as well as to show interest and attention, to ensure that any disturbing information will reach the project leader early and in a non-threatening manner.

In the Fable, Simon is successful in weaving a network that gives him and advance warning of the mussels issue; and allows him to connect effectively with that segment of his stakeholders. On the other hand Harry fails completely in that respect, and is completely overwhelmed by this potential show-stopper.

A word on the usage of communication professionals to do the 'Public Relations' bit ('PR people'). They can help, they can advise, they can produce useful material, and they can give feedback about perception, but they cannot replace the direct, genuine contact between the project leader and the key stakeholders—and they should not decide what the overall strategy (the 'content') is. The intent is that the key stakeholders should be able to grab their phone and reach the project leader or vice-versa, based on the level of confidence between them. No proxy can replace this, and the successful project leader has to establish and nurture the key contacts himself, demonstrating his authenticity during the process. Also, be wary about advice from PR people who are sometimes nervous about or scared by potentially conflicting situations (often linked to their fear of professional failure), and who will do what they can to avoid those situations. When managed properly and boldly, these conflict-laden situations can be decisive turning points for the success of a project.

Nurturing the network

Everybody knows it, but not many people practice it consistently: once established, a network needs to be continuously nurtured.

It is critical to be able to nurture the network in the most effective way possible, using the least time necessary (as the project leader obviously has other activities to attend to). There are several techniques to achieve this objective:

- the effective network needs to be limited in scale to be effectively nurtured, with quality time devoted to the relationships therein (for more information, refer to the guidance given on the topic of network creation);

- nurturing needs to happen in a differentiated manner which varies depending on the criticality of the network's members, with more time being spent on the most critical members;
- maximize the scalability factor. Modern technologies allow project leaders to leverage newsletters, dedicated emails, webcasts, videos, and other communication tools. Scalability can also be achieved through specific meetings, visits, etc.;
- use the technologies of the Fourth Revolution to create an online community where the network can interact while the project leader minimizes the time spent producing content by himself. The ideal is to create a community that sustains the conversation by itself, independent of or with only minimal input from the project leader.

The key is to keep the network engaged by continuing to follow the two golden rules of giving and listening. The successful project leader will ensure that key network members are engaged consistently, personally, and with an established frequency during the project. Dashboards following up these contacts can help ensure that they happen consistently and regularly.

Maintaining the connection against all odds

Sometimes situations occur where the relationship becomes tense with some key stakeholders. It might be because of events that are currently happening, because of a particular negotiation trick, or because of the relationship style and character of the persons involved. The successful project leader knows that he needs to maintain the connection despite all odds. He should not give in to the easier path of reacting instinctively and breaking the connection. Should it happen, the effort spent to get back on track will swamp the emotional and physical energy of the project leader for most of the rest of the project, impeding concentration on the actual project delivery.

Of course, it is not easy to remain calm and composed in front of an abusive, angry, or loud stakeholder, in particular when the stakeholder happens to be the client of the project! It is not easy to stay upright and defend one's field calmly while being insulted or otherwise threatened; it is still harder to do this while maintaining the connection. Yet that is both necessary and vital.

The successful project leader recognizes that maintaining the connection against all odds is necessary to the project's success and will overcome his naturally defensive feelings to keep in touch with all of the key stakeholders. Self-awareness and self-control are key practices during these occasions which can make or break a project.

The same issue can arise during negotiations. It is always important to separate the person from his or her role in the negotiation. While maintaining a clear objective for the negotiation, it is vital to establish and subsequently keep the connection with the other party despite the odds.

Using the network

Once the network has been solidly established and continuously nurtured, which involves both continually engaging and listening, the network can be used, i.e. the project leader can start benefiting from it.

Ideally, most of the network's usage should be passive for the project leader. An effective network will give back proportionally to what it has received, and return many times more value to the project leader than what he has given or invested.

Advanced uses of the network to actively influence a group of people need to be treated cautiously. The right level of rapport needs to be achieved first, and the project leader must be careful at all times not to compromise its key network members by making them appear to be too friendly to the project. Still, once the right level of interaction has been achieved, the network can be leveraged in incredible ways that allow the project leader to reach large areas of the project's environment.

Going into the field

The project team—or, more broadly, the people who are involved in the actual execution of the project—is an important part of the project network. They are absolutely key stakeholders and are vital for the effective delivery of the project! In addition they are on the shop floor, the fingers deep in the grease or the mud. They need to be listened to, because they often know what is happening better than management does, and they are often very experienced within their specialization.

Time spent in the field listening and establishing rapport is not wasted. It is a must for the successful project leader. Once solidly grounded in the realities of the field, and having himself felt the mental and physical difficulties faced by the crews, the project leader can make more applicable and appropriate decisions. He can also identify upcoming difficulties more easily.

A very fine way for the project leader to gain goodwill in the field is to identify one or two niggling issues that are frustrating the team—these often revolve around access or quality of toilets, drinks or food, or frustration caused by irrelevant bureaucratic rules—and to take it upon himself to solve the issue immediately. This demonstrates concern for the people toiling on the ground and can also improve morale. Still, this does not replace the need for deeper rapport with a limited number of key individuals in the field.

By field, we include of course the main project activity sites, but we also mean the main subcontractors' and suppliers' worksites. When they are key for the project's success, then they should be considered part of the key stakeholders and they therefore need to be included in the project leader's network.

Many project managers effectively avoid going into the field. It is not for the reasons they claim when they say they have had no time, because of more pressing issues. This is just an excuse, because it means that going into the field was not high on their list of priorities. Often it falls to the bottom of their to-do list because they are afraid of facing the reality of work in the field. Field visits can be draining moments, more so if the project leader does not go very often, because he will be faced quite suddenly with the inadequacies of what has been imagined by those engineers and other project staff who are comfortably installed in a nice office. That is another reason to maintain a discipline of going into the field regularly, as it ensures that one's vision of reality is up-to-date, and allows the project manager to tackle any inadequacy in working conditions rapidly.

In the fable, Simon goes one step further. He arranges for his office to be close to the work site—in sharp contrast to Harry, who keeps his office in the capital city—and Simon also uses the opportunity to mesh office and site people into a single team.

The internal network: leveraging both formal and informal organizations

In addition to going into the field, it is vital for the project leader to get candid feedback about what is happening in the project at any given time, and this is even more important if the project team is large.

To achieve this, the project leader relies on a formal organization with direct reports. That alone should be sufficient. Still, the project leader also needs to check the alignment of the informal organization.

This is why Simon is running periodic mood checks of the team through anonymous questionnaires. There is another way to connect to the informal organization: create some kind of periodic or continuous activity that cuts across the traditional hierarchy and allows the project leader to effectively establish rapport with a cross section of the project team. Of course the successful project leader will not fall into the trap of letting this rapport influence his management decisions or his judgment regarding these people, and will leverage this rapport for the benefit of getting a candid view of what is happening across all levels.

What activities can be used this way? They vary across cultures and habits. It can include periodic informal dinners, parties, or sports games. While not recommended for the sake of one's health, smoking is also a very effective way to establish rapport across social barriers, since smokers are forced to squeeze into crowded spaces outside buildings in order to indulge their habit. No matter what activity is used, the key is to spend time in a relaxed environment with a cross section of the project team.

The project leader needs to identify key influencers in the team who can effectively leverage the informal organization. They will become obvious over time. Developing rapport with these people is key to getting timely feedback from the informal organization, and may serve as a way to pass some messages through, which is sometimes more effective than using the formal organization. The successful project leader knows when and how to use the formal or the informal organization to communicate with the members of a project team.

Particular issues of multiple project offices

Large projects might require the project team to be located in several project offices that are located thousands of kilometers apart and which may be subject to large time differences. In these cases, a key concern of the successful project leader is to keep the internal network operating at the right level. The project leader and his core team have to spend a significant portion of their time in making sure the internal network works well. A significant investment is required to make sure that people know each other well enough so that they can work reliably and efficiently even in remote offices. The internal network really demonstrates its importance in that case, as well as showing how the interfaces between each project office are structured.

The Art of Disbanding the Network

While the project network is primarily established for the purpose of facilitating project delivery, it should not just be abandoned at the end of the project. Projects induce change, and change needs to be sustained. The new environment created by the project deliverable will create a new dynamic. The project network is the basis for this new environment.

It would be a shame to ruin months of effort spent in creating a sound network of key stakeholders by not properly managing the transition to the new situation.

The successful project leader spends adequate time and effort in celebrating the project's completion with the project network, and works to generate the basis for an ongoing network which will continue on in the future. There are a number of options available. Some people will leave, and the network composition might change; also, the project leader might or might not stay involved. In any case, it is the responsibility of the project leader to set the project network on the road to success, and to leave everybody with fond memories about the project. This is why Simon organizes an after-project network.

People always remember the beginning and the end of a story, but not so much the middle. What will be remembered of a project are, in large part, the events that take place at its completion, such as the awe of the clients, the team's celebration of its success, and the positive celebration of success by the project network. The successful project leader ensures that everybody, and in particular key members of the project network, remembers the project in the best possible light.

Summary

The Spider's role, that of weaving and maintaining the project leader's network, is an essential practice. It needs to be made effective by developing and nurturing a specifically targeted network related to the project. It is also important not to forget the internal network. Going into the field and leveraging both the formal and the informal organization are key practices. These activities require proper planning as well as the successful project leader's personal attention and time in order to develop strong personal connections with relevant people.

Second Role: the KUNG FU MASTER, Maintaining Discipline and Focus

"Focusing is about saying no. And the result of that focus is going to be some really great products where the total is much greater than the sum of the parts" – Steve Jobs, 1997

Like the Kung Fu Master, the successful project leader can practice the disciplines of project leadership for hours, days, months, and even years. He can focus intently on a subject for hours, silent and patient, and then strike like lightning, suddenly shattering the seemingly unbreakable obstacle with a focused strike of his bare hands on its rough surface.

Discipline is a key attribute of the successful project leader. It means relentlessly practicing the fundamental disciplines of project management, making sure the underlying processes deliver week after week, month after month, with the most limited variance possible.

Focus is another key tool of the project leader. Success is the result of the project leader focusing his attention intently on the few important things that drive the project.

Discipline and focus are fundamental to outstanding execution.

The discipline of project management

Successful project leaders know the value of the discipline of project management. They know the importance of applying fundamental processes in a rigorous and effective manner. A list of these vital processes can be found in any basic project management course. For example, the Project Management Institute's Body of Knowledge identifies the following main processes: integration management, scope / time / cost management, quality management, human resources management, communications management, risk management, and procurement management. In addition, Health & Safety management is often a key process for large infrastructure projects .

Why are these processes so important to maintain? Because the project leader will have an accurate view of what is happening on the project if, and only if, these processes work appropriately. Should these processes fail to function correctly, the project leader will be like an aircraft pilot flying through a storm at night without reliable instruments. A crash of some kind is almost certain—and there won't be many survivors!

The Project Soft Power leader is not just a practitioner of a new touchy-feely sort of project management style. The successful project leader knows that effective discipline is crucial. Discipline as in a repeated, consistent exercise that is maintained on the critical project management processes day after day, month after month, no matter what else happens.

The successful project leader makes sure that the planned review meetings happen, and that the right reports —those that effectively give a true insight into the project's progress—are being produced, analyzed and discussed. He makes sure that people know the key rules that make the processes reliable, and ensures that people follow those rules. The successful project leader makes sure the entire team practices the required disciplines and exercises consistently. In the fable, Simon makes sure that the regular project review meetings happen like clockwork and that the most important issues get discussed; by contrast,

Harry just relies on his indicators and is not as worried about the basic disciplines of project management.

Like the Kung Fu Master, the successful project leader implements an exercise routine, practicing the project processes day-in and day-out, whether the weather is fair or poor. The successful project leader makes sure that even during the strongest hardships, the key project management processes still get the priority and attention they need.

This does not mean that the project leader does not have time to devote to anything else; he will have established an organized system so that the basic project processes can do their work and produce their results with minimal involvement from the project leader. Still, he will pay sufficient 0attention to these processes to make sure they work well; and if they don't, he will work quickly to get them fixed. He will regularly receive a limited number of hand-picked indicators that will keep him aware of the current status of the project, and he will make sure that he devotes sufficient time to the task of analyzing them.

Identifying the focus areas: the theory of constraints, applied one constraint at a time

The theory of constraints, a famous empirical theory developed by Dr. Eliyahu Goldratt, basically states that at any moment in time, one single constraint (or, very rarely, two) drives the outcome of any endeavor.

That would be a simple enough theory to grasp if that constraint were always obvious. But this is not always the case, because the vital constraint might have no immediate symptoms. For example, the constraint might be linked to a future activity—perhaps one which will bring the entire project to a grinding stop in a few months' time although nobody cares about it now—or it might be an underlying cause with few observable symptoms.

Still, in our observation, the constraints theory holds true. At any given time in a project, there is one single constraint that unambiguously drives the project's delivery.

It might not be easily identified, but it will reveal itself eventually.

The successful project leader knows how to identify this constraint early enough by carefully observing the project's development. And once the constraint has been identified, the successful project leader knows how to effectively concentrate his effort in order to overcome the constraint.

In the fable, Simon repeatedly focuses his attention on the few constraints he believes will drive the delivery of the project. He applies this practice to the point where he does not show much concern about the progress of the most visible piece of work, the digging of the shaft, because at that stage he already knows that it is not the main constraint.

Identifying and acting on the project constraints

In Goldratt's classical approach, which came from the world of manufacturing and therefore centers on manufacturing-related processes, constraints could be somewhat easily identified by the length of the queue in front of the constraint, or work-in-progress—like the height of a pile of papers or a stack of semi-finished products. These identification criteria do not really work in a project-centered environment, because it is a transient environment that does not involve repetitive tasks that take place regularly over long periods of time, and thus does not allow the buildup of inventory.

Project management scholars have been struggling for a long time to come up with a simple way to identify constraints in a project-centered environment. An entire field of project management theory, the Critical Chain Project Management, or CCPM, has been developed around the theory of constraints applied to project management. This method entails a number of real breakthroughs in terms of effective project management, but it remains complex and so the number of people that really understand it is very limited. Also, while it improves the project manager's ability to predict the overall project's pace, it does not necessarily prevent projects from failing completely. It is slowly gaining ground as CCPM analysis software comes onto the market. Still, it has not gained widespread acceptance, probably because it is difficult to visualize how CCPM works.

CCPM teaches us one fundamental insight: resources are the fundamental constraint that drives overall project pace. Actually, some key resources, very limited in number, drive the overall schedule of the project, and do so independently of the conventional critical path. This aspect is almost completely overlooked in conventional project scheduling, because of the basic Industrial Age mindset, which holds that resources, and in particular, human resources, are exchangeable and can be easily brought in and removed. While resource leveling is a way to address this, the way it is being done by today's software is arbitrary and thus does not necessarily reflect the best real-world solution. And in our experience, resource leveling is not carried out in large, complex projects because it would be too cumbersome to implement.

So, the fundamental insight from CCPM is that some key resources are almost always a driver. In real life, people have particular talents and are not easily exchangeable; they cannot be moved around tasks without significant loss of productivity, and their actual productivity is rarely measurable in norms, particularly where creative activities are concerned. And specific individuals can have an effectiveness that is orders of magnitude higher than the effectiveness of most others. In addition, some teams can demonstrate levels of effectiveness that are orders of magnitude higher than those of other teams.

Another classical observation is that a project can generally be described as a set of activities converging toward a single outcome. As such, the meetings of deliverables at the convergence points must happen in a timely fashion. Failure for these activities to converge appropriately explains why large projects can fail or have major cost overruns because of a relatively minor mishap or forgotten element that holds back the entire project until a solution is found. This occurs because time costs money, and a project team with all of its associated equipment cannot easily be disbanded and remobilized later. Furthermore, delays caused by minor issues can cost enormous amounts of money at the global project level. As such, relatively minor or cheap elements can sometimes become the constraint that drives the entire project, and it is entirely possible for a missing screw worth two dollars to cost hundreds of thousands of dollars or even millions of dollars in direct and indirect costs due to resulting delays.

In the fable, this is what happens to the Norlanders. At one point, they discover that they are missing the 'locking disks,' relatively simple components whose absence brings progress to a standstill throughout the entire worksite. The effort of making sure these components become available involves mobilizing a great deal of effort and energy, distracting the project team from other important tasks.

The tools that are traditionally utilized to prevent this from happening are called "convergence plans," a concept invented by Toyota in the 1970's. In convergence plans, key convergence points of the project are identified, as well as the most crucial deliverables. Dates are set, and a disciplined system is implemented, in which any slippage of these critical deliverables is unacceptable. Should there be delays, management will commit to providing the resources necessary for a full recovery. The rationale for this tool is clear: convergence points even in remote branches of the project tree cannot be late, because otherwise all the rest of the project will be impeded by it.

Let us now put these two insights together. A few key resources, and in particular human resources (which are more difficult to substitute), generally drive the project's overall pace and schedule, and small items that go missing at critical convergence points of the project can unexpectedly stop the project from progressing, thereby causing significant delays and increasing costs.

The constraints that a project leader has to care about are therefore of two main types:

- the few key resources (most often, key people) who will drive the overall pace of the project because they are key to many simultaneous activities
- the key deliverables that might be missed at project convergence points whose delivery must be monitored so as to anticipate and prevent unfortunate delays.

In both cases, the following criteria can be used to screen the key resources and deliverables that could qualify as constraints:

- they are difficult to substitute, perhaps because the resources are expensive, take time or skill to make, or are unique and cannot be substituted, period;
- they are scarce and are highly utilized throughout the project or throughout a major project phase;
- they are difficult to produce, perhaps because they are new and unproven, or because their production is not under the project team's control, or because they already depend upon the convergence of many other deliverables, thereby increasing the risk of delays.

Once a constraint has been identified, the response is quite simple:

- **Support** the constraint, by all possible means. Ensure that all the other resources support the key resource or are involved in producing the key deliverable, In particular, ensure that the key resource is focused on producing what is expected with no distractions and without any need to multi-task. Others can literally stop doing what they normally would do for a while, because the project is going to stall anyway if they don't support the key resource.

In addition, in the case of key resources, the following approaches can work:

- Short term: **Focus** on the constraint. Make sure the key resource is busy as it can be by removing unnecessary tasks and by ensuring that it is focused on the delivery of the project.

- Long term (often impractical in a project unless anticipated well in advance): **Elevate** the constraint. Increase the capacity of the key resource through investment or recruitment.

Focusing on the constraints: saying no

The life of the project leader is full of assorted solicitations. Some (unsuccessful) project leaders let themselves be driven by these solicitations, and try to respond to each one, thereby dispersing their energy and their effort and ultimately achieving nothing.

The successful project leader, after having identified the constraint, first engages in one practice: **saying "no."**

The correct practice is to say no to a large percentage of unnecessary solicitations. This is particularly difficult because the project leader often has the obligation of representing the project if the public or the media wants to know more. Also, a project leader has many important tasks that need to be attended to, not the least being the practices advocated here as part of Project Soft Power. Simon in the fable says "no" even if it is difficult for him. On the other hand, Harry is permanently unfocused, and travels the world giving speeches and interviews. These different behaviors create very distinct impacts on their respective teams.

We are not suggesting that project leaders should spend all their time focusing on the constraints; rather, we are advocating that the constraints should have top priority, and that the project leader should spend a sizeable portion of her time, at least 30%, on those constraints.

Furthermore, those constraints should significantly inform many aspects of the project leader's activities. For example, communication should be significantly influenced by the constraints so that project team and network alike are aware of the project's top priorities. Decision-making should be influenced by whether an option helps to solve the constraints, and so on and so forth.

Project leaders often blame their lack of available time on external causes. But we have often found that ineffective project leaders fall into the trap of indulging in personal practices that burn their time.

Let's face it: addressing the project constraints is hard. It is hard because project constraints are the key to the project's success, and the project leader is there to solve this difficult issue with the support of his team. (Remember, if there were no constraint, there would be no project or project leader, because anybody could complete the task.)

And, when faced with difficulties, most people procrastinate. They will find all sorts of excuses to avoid dealing with the issue in front of them. All of us do this, to some degree. Preferred personal procrastination methods can involve, for example, aimlessly surfing on the Internet, sitting passively in meetings all day just because our presence there is well-thought-of, spending hours on phone calls where a few minutes would have been enough … just to mention a few.

The successful project leader boasts a strong personal discipline mixed with the capability to say no. Say "no" to sitting passively in meetings that bring nothing to the project. Concentrate your efforts instead on the few key practices of Project Soft Power, which will help to make you more effective. Drop the rest.

Once you have focused on the constraint, execute with discipline by relentlessly following through

Once the successful project leader says "no" to the activities that are not really important for the project's success, he can focus his attention and his energy on dealing with the project constraints and the other few practices that are essential to project success. Focusing time and attention means *quality* time and *full* attention, which will necessarily promote a more prompt resolution.

Still, some unsuccessful project leaders fail at this stage, and what differentiates successful project leaders from unsuccessful ones is the implementation of a specific practice: **relentless follow-through and discipline.**

It is not enough to say "no" and to focus on priorities. Effective, consistent and publicized follow-through on these priorities is the mark of the successful leader. It shows everybody that yesterday's priority is still a priority today, and ensures that priorities are taken seriously.

It is quite uncommon in organizations to see leaders effectively follow through on a priority they declared several days before, because most leaders allow themselves to be pushed around by circumstance. The day a leader effectively follows through and remains steady and consistent with his priorities, people notice. And soon, the other members of the project team will follow up, too.

The essential quality of top-notch discipline is that it ensures consistent action over time, despite whatever else happens to the project. Effective follow-through requires repeated, regular follow-up at fixed intervals. Others may resent being asked repeatedly about actions that they haven't managed to make progress on. If those actions are decisive for the project, they *must* happen; therefore the successful project leader does not diminish the pressure for delivery and makes sure that the right resources are brought in so that the issue is solved in a timely manner.

One of the most important areas where a project leader needs to be consistent is on the projected schedule for priority activities. Let's say something needs to be done in six months' time. and it is a project constraint, the top priority of the project leader. At the next monthly meeting, it needs to be made very clear, and widely communicated, that the objective is now five months away. That will catch people's attention, and ensure that instead of having a team lose track of time and think that the objective is still six months away, the team will pay attention and be more prepared. In a project, the schedule is absolutely critical, and relentless follow-through must focus on time and take into account the progress that has already been made.

Once priorities are clear and limited in number, it is quite easy for them to become a fixed preoccupation, something team members are aware of and will monitor on their own to some extent. At that point, a small follow-up board is enough to keep track of progress regarding the very few main priorities of the project leader.

The trap of the details and of the analysis

Because project leaders often come from technical functions where precision and attention to details was a fundamental evaluation criterion in the first years of their career, they often tend to keep that bias toward copious detail even after they are promoted. "Focus on detail" is a contradictory sentence, yet it can easily become a trap for the inexperienced project leader. An excessively analytical approach is a close cousin to this problem. In the fable, Professor Hugh's theoretical approach leads him to impose an excessively detailed plan on the Norlanders, which leads to disastrous effects when the plan could not be updated reliably. Let us dwell a moment on that extremely common disease amongst project leaders, particularly junior ones.

Excessively analytical project leaders will spend long hours developing detailed models of their projects, considering their projects in terms of cost, schedule or other ways of planning. Their schedule will typically involve extreme detail. Software packages today allow leaders to construct an almost infinitesimal breakdown of any project. But just because a leader can do this, does not mean that a leader *should* do it!

This is a trap that will catch the project leader sooner or later, because of two main issues:
- the project plan/schedule will be overly onerous to maintain and update, wasting the time of project team members on repetitive, mechanical tasks, and preventing them from looking at the big picture;
- the project plan will be developed in such minute detail that it becomes difficult or impossible to identify the main drivers (constraints) of the project.

We have seen this analytical tendency become a real disease when project leaders spend long hours and longer evenings developing detailed plans. In the fable, it is a disease that plagues the Norlanders' team, leading to a final breakdown of their management systems and thereby preventing them from having a clear view of what was happening at the most critical moment.

Dwight Eisenhower famously used to say, "In preparing for battle, I have always found that plans are useless, but planning is indispensable." Things will not happen exactly according to your plan. Sh*t will happen. Bad luck will strike the project. So don't overplan; that is simply a waste of time, because the detailed plan will end up being useless, anyway. Instead, identify the drivers, plan in detail the few bits of the project that are really critical, and stay at a relatively low level of detail when planning out the rest. This will minimize the amount of data that needs to be shuffled around every reporting period, and reduce the extent to which your plan will need to be reworked when things inevitably change.

Analytical minds often spend inordinate amounts of time developing the most precise project plan they can manage. In their eyes, it's precise, but let's face it: they probably do that partly to reassure themselves, to avoid confronting reality, which is that things rarely if ever happen they way they did in the plan.

By spending too much time on planning and by exhausting their team by forcing them to deal with excessive amounts of nigh-worthless detail—to say nothing of the long hours wasted in crafting categories for each of those details—unsuccessful project managers do not spend their time effectively. By contrast, the successful project leader adapts the level of detail to the criticality of the issue, makes sure the schedule reflects a desirable balance between the different types of activities, and most importantly, identifies the constraints and focuses on them once the constraints have been found.

The practice of the 'two plus one' rule

The 'rule of two plus one focus' followed by Simon in the fable is a direct application of the practice of focus. As explained by Simon, the rule is to have a maximum of two objectives that related to the project—one short-term objective that can be achieved within one to two months, and one longer-term objective that can be achieved within six months. (These durations are guidance for projects spanning over more than one year. For much shorter projects, shorter durations should be used; as a rough guide the short-term goal should change 8 to 12 times during a project and the long-term goal should change 3 to 4 times.) In addition, the project leader always has a particular goal that is related to the health of the team itself.

In addition to setting goals, the rule of 'two plus one' publicizes what the project leader is currently focusing on at the moment. This has two benefits:

- By making a commitment publicly, the project leader ensures that the entire team will provide feedback to him regarding any deviation from his self-imposed focus, thereby triggering a self-correcting mechanism—self-imposed social pressure being an extraordinary support for continuous, sustained effort;
- The team knows at any moment what the project leader is focusing on, and will understand accordingly the reasoning behind some of the project leader's actions. The team may also help work toward the resolution of the problem if, by any chance, any of the team members can find a way to do so.

Four common questions about discipline and focus

When exposed to the practices of the Kung Fu Master, the concept of discipline and focus, project practitioners often come up with questions which revolve around the four following issues:

Q: You advocate concentrating on one or two constraints. What if, as a consequence of this, I don't pay enough attention to something that turns out to be important or even vital to the success of the project?

A: Excellent question. And the fear of not addressing something important is often paralyzing to unsuccessful project leaders, who then do not properly prioritize.

Prioritizing effectively means letting go of the things that are not currently a priority. It does not mean that a project leader should do everything else, and then on top of those duties, somehow do more of what's really important. No; that leads to lower effectiveness and higher stress. People who practice that version generally divorce at an early age and are sadly at risk of dying from a cardiac arrest way before they can enjoy their retirement.

The successful project leader needs to let go of what he believes is not so important. (Note that this does not mean these tasks do not need to be done; it just means that these tasks do not warrant the full focus and attention of the project leader.) How can he do that safely?

- First, he spends sufficient time understanding the drivers of the project and going through the process of determining the project constraints. He can't be too far off.
- Second, even if for some reason his identification of the real project constraint was flawed at the time, through his project network, his team, his convergence planning, and similar early warning systems, he can identify in advance that there is an even larger constraint coming up to block the way. This affords him the time necessary to rearrange his priorities.

The basic message is this: if you don't have clear priorities, you will spread your energy around until your effectiveness is close to zero. If you want to really go somewhere with your projects, you need to focus. Take some time to analyze the situation and to focus on what appears to be most useful. You'll make progress. If it turns out that there is another, bigger constraint that was hidden at first, be flexible and take this into account. You probably see it now that you have made some progress anyway, so just deal with it. Priorities do change. That's life. They should not change too often. And they should be real priorities.

Q: You say that dealing with project constraints is hard work. Does that mean I need to work twenty hours a day as a project leader?

A: No; quite on the contrary. The successful project leader is not an overstressed person working twenty hours a day. Granted, there might be some tough moments in a project, like when a new system or installation goes live, which briefly necessitate very long working hours. Still, these moments should remain reasonably rare.

Because the successful project leader is saying "no" as a consistent practice, he has much more time than other less successful colleagues do. In fact, the successful project leader needs to have time to solve the difficult issue of the project constraints, and for this, he needs to have time to discuss, network, reflect, research, strategize, and so on. The successful project leader will thus look engaged and focused, and should certainly not be overworked and burned out. The successful project leader is a balanced person who maintains an open mind while leaving sufficient time for other activities.

Q: You speak about maintaining a strict discipline of project management processes at the same time that the project leader needs to concentrate on a very limited number of constraints. Isn't that contradictory?

A: It might look contradictory at first. But it is only with a proper measurement system, one that gives a clear indication of the project's current situation, that the project leader can accurately assess where the constraint—the top priority—is. If the project management processes are broken, fixing them should be the first constraint that needs to be resolved, before the project leader even starts seriously thinking about the project deliverable. It is not sustainable for a pilot to continue to fly his aircraft if the instruments are known to be wrong. As soon as a storm hits and visibility falls to zero, accurate instruments will be the only indications of what is happening.

As in any organizational software, processes are maintained in a functional state through consistent discipline and execution. The successful project leader spends the necessary time, in particular at the beginning of the project, to make sure his instruments give the right indications; and he must not forget to recalibrate them from time to time.

Q: Does focusing on project constraints mean I need to drop all social activities, like discussing the project with people, or having coffee with them?

A: No, not at all; quite the contrary. This is why: The project has generally not been done before, or not under these conditions. (This is why it is a one-off project.) Some solution to a difficult problem needs to be found in order to allow the project to progress. Finding and relentlessly implementing a way to solve the main project constraints is thus a creative act.

As a creative act, it is about finding ideas. The idea of the lone genius in his ivory tower is long obsolete. We now know that the creative process occurs by rubbing ideas on each other, finding similar issues in widely different fields. In summary, it involves an open mind and an intense sharing and combining of idea. To do that, the successful project leader needs to enhance the probability of encountering the right idea that will allow him to solve his constraint. He needs to be on the road and moving!

Taking a break for coffee can become an act of procrastination if extended, but it can also become the most useful moment of the day if the conversation during the break is focused and the participants' minds are open.

So, focusing on project constraints effectively means spending more time in effective inter-personal relationships. Surprise—that's what most Project Soft Power practices are about!

Summary

The Kung Fu Master is all about discipline and focus. A relentless discipline has to do with running the basic project management processes. A relentless focus deals with choosing priorities and effectively sticking to them.

A project will be driven by a very limited number of constraints, which are often resources or deliverables for critical convergence points. The key role of the successful project leader is to identify these constraints and to focus on overcoming them. Only by overcoming the project constraints through habits of relentless focus will the project be delivered successfully.

Third Role: the ENTREPRENEUR, Building the Project toward a Purpose

This might at first seem contradictory, because a project is fundamentally a short-term endeavor: still, the successful project leader is fundamentally an entrepreneur, because he builds something, even if it is only temporary. He builds his project, much like a vehicle, to reach certain objectives. Obsessed by his objectives, the project leader is able to postpone short-term gains for the sake of long-term benefits; he is able to take reasonable risks.

The Entrepreneur mindset is often the key that distinguishes between the successful project leader and the average project manager.

A really successful project will necessarily go through painful phases

If a project goes smoothly, without any difficulties, then the project was probably not challenging enough for the project team. Maybe luck played a part, but the organization was probably under-ambitious with regard to that particular endeavor. Moreover, talent was probably wasted as a result.

The bottom line is, every project that produces something significant, something that really changes its environment, will go through some difficult phases—times where challenging decisions need to be made. This will occur because the project is trying to create something never previously achieved, or at least never achieved in that location or under those circumstances. Or else the project is trying to create something so totally new to its stakeholders that a simple, natural resistance to change will create all manner of obstacles. Or, perhaps the schedule was challenging, or there were other project constraints that were unidentified.

The successful project leader makes a difference because he is able to navigate the rapids that mark the sometimes-unexpected course of the metaphorical project river. During these times, a leader is decisive, anticipatory, living his long-term vision. He stands at the stern of the ship, his hand on the tiller, his eyes concentrating on the distance, keeping course. He is an Entrepreneur in his purest form. He is a real leader, respected by the project team.

During these harsh times, project leaders will be confronted with difficult decisions. They will often have several paths available to them. Some of these will be easy to manage in the short term, but bear the seeds of a lousy long term; some other options will be difficult at best in the short term, but bear the seeds of a bright future. The successful project leader decisively embarks upon the path that will deliver the best long-term results, even at the price of suffering in the short term.

The obsession with the outcome

And indeed, an entrepreneurial mindset is a fundamental characteristic of a successful project leader, because what he will be judged upon, fundamentally, is whether he was successful at the project's completion. The end of the project is what counts. How the successful project leader attained that final outcome is the result of the entrepreneurial project leader's workmanship.

The successful project leader's attention is thus concentrated on the results at the time of project completion, and on whether the results are largely positive in terms of expected outcomes, whether financial, organizational, or social. To achieve significant change and remarkable results often takes a lot of preliminary investment, pain and frustration. It often takes a lot to overcome resistance and obstacles. Persistence is the key.

In the fable, Simon is constantly focusing on what will give him good results over the long term. This leads him to making decisions that are sometimes not so great in the short term, as in the case of the shape of the shaft. On the other hand, Harry's team is focused on optimizing the present situation without due consideration of the future benefits or drawbacks of that decision. And this makes a significant difference when the going gets tough!

Moreover, it is amazing to observe how successful project leaders shape much more than the project outcome. They will often foster and develop great individuals on their team, influencing the organization in the long term. Their investments, implemented as part of the project execution—whether they be investments in infrastructure, equipment, or systems—are often worthwhile building blocks that can be reused later, benefiting people, even after those investments have already paid for themselves within the project by making the project team's task easier.

About the importance of determining the expected outcome of the project in detail

The previous discussion is based upon the premise that the project leader has a clear understanding of the expected project purpose and of the expected outcomes. It might seem odd, but our experience shows that project goals are not often clarified enough at the onset of the project. While the overarching goal of the project is often obvious—build this, make a profit from that—it vital to take the time to define the project goals in detail, taking into account the stakeholders' expectations, and involving the core project team.

The importance of this process of setting project goals should not be underestimated. Here are some recommendations:

- The development of the project goals should be based on stakeholder analysis, so as to ensure that no important stakeholders are forgotten. The project team's needs will be addressed, as they, too, are part of the project's stakeholders.
- The project goals should be publicized to the team and to the main stakeholders in order to promote about a clear understanding of both the expected outcome, and the direction in which the project leader is steering the project team.
- Project goals should be broken down into project objectives with a shorter duration, particularly for long projects, by identifying specific objectives by year or by quarter in the framework of the overall project goals.

Resisting external influences to achieve the project outcome

While the successful project leader is focused on the project outcome, he will be continuously distracted by a large number of external influences. And during hard times, when the project's success can more easily be called into question by stakeholders, he will be under tremendous pressure.

The most important task for the project leader in situations like this is to gain time and space so as to be more effectively able to execute the project. Unsuccessful project leaders will fall into the trap of spending too much time managing their stakeholders, removing their attention from the project execution. In the fable, Simon is able to say "no" to solicitations and to Albert's attempts to influence, knowing that this is important for the sake of the project. On the other hand, Harry, who is driven by his ego, does not always put the project outcome first when making decisions.

This is where the network developed by the project leader becomes particularly handy. The successful project leader knows how to leverage this network to minimize distractions. He will use some supportive stakeholders to respond to other, more aggressive ones. A different tactic is to delegate part of the response to external stakeholders to other members of the project team. This could go as far as placing a dedicated member of the project core team in charge of managing these relationships, if the situation warrants it.

At the organizational level, this issue also has consequences. "Departmentalized" or "weak matrix" organizations, where the project leader has only a limited influence on the composition and operation of his project team , are not favorable to disruptive projects. Real game-changing projects need to be executed in an organizational setup where the project leader has the leverage to make appropriate decisions in terms of financial investment and personnel management. The project team is then seated together, somewhat separate from the rest of the organization, to reduce distractions.

Resolving the conundrum of planning and preparation: the ROI practice

At the outset of a project, the project leader is faced with a whole set of structural decisions having to do with the project execution.

Faced with these tough decisions, the successful project leader engages consistently in the following practice: **What's the Return on Investment (ROI)?**

This practice is difficult, because it means that obtaining results requires an investment. An investment upfront involves a possibility of not seeing anything worthwhile in return. It is thus risky, and requires short-term sacrifice. This is why it is a conundrum for the inexperienced project leader. If he is scared to make the necessary investments for fear of being unable to justify the expenses, he will later suffer from inadequate support or perhaps face the impossibility of completing the project.

This practice is also made more difficult by the fact that it is not already in the DNA of most traditional organizations. In these "Industrial Age" organizations, managers are intently focused on costs. In their view, costs need to be kept minimal, and ROI is not a concept that they understand very well. The concept of spending money now for a possible future return does not easily permeate middle management ranks. The successful project leader is not just cost-minded; he knows that good results require investment up front, and he is willing and able to make the decision to spend money now for the sake of receiving future benefits later. The fact that the project leader is focused on the final project forecast outcome, and not on the current month's cost, is key to overcoming the cost mindset.

From our observation of the practice of project management, some particular actions always have a very large ROI when implemented at the beginning of a project. Still, they are not always consistently pursued by the project leaders, sometimes with very serious consequences:

- **The project controls setup phase**. It should normally involve a number of weeks of excruciating hard work at the very beginning of the project to set up the navigational instruments, i.e., organize the project reporting and measuring and the relationships/interfaces with external entities with regard to resource utilization and reporting. When done properly, with the right level of investment in organizing and structuring the information, this will allow the project leader to have access to accurate measurements of performance and activity with little effort. If this is done poorly, the project leader will be in the dark during project execution, unable to be certain about his project performance, and will spend inordinate amounts of time trying to sort out issues with unhappy stakeholders communicating about unexpected surprises that have affected project performance;

- **The systems and processes development phase**. Its intensity will depend on the maturity of the organization; still, there are always some systems and processes that need to be put in place for a particular project. Setting them up will take time and require the involvement of the project leader. These critical pieces need to be identified early, and the investment made as soon as possible, to make sure that they will be available when the project gets going;

- **The office setup phase**. Particularly in the case of long-haul projects, it is important to give the project team a home they can feel comfortable with. This might necessitate some investments in the areas of office renovation and equipment, or even require the creation of an entire project office. If it is possible to be flexible on the matter, the physical arrangement of the office is a fundamental factor in enhancing collaboration. For example, partitions between desks in the open space need to be lowered so that people can discuss issues with one another without having to stand up, and space needs to reserved for collaborative work on documents, while white boards need to be available for brainstorming, etc.

- **The project team setup phase**. It will requires some investment in time and resources to ensure that the project team members get to know each other a bit more intimately than the usual superficial working level. It is also important that the team and sub-teams work effectively together, and that they are balanced and sufficiently provided for in terms of their talents and skills. A project leader will spend a large percentage of his time choosing, integrating and engaging project team members at the onset, building the right atmosphere and promoting good practices. Shortcutting this investment will expose the project leader to poor overall performance, and worse, a team that avoids responsibility at the most critical moments of the project. (At that point, the project leader will turn around to check that he has the backing of a strong team, only to discover that nobody on the team has followed him there!);

- **The phase in which a project dedicated network is built**. This is the process that is described in detail in the 'Spider' role.

In the fable, Simon knows from experience exactly how important it is to invest in these areas, and he does so upfront.

Why the conventional approach to "risk management" is inadequate

Risk management is a process used to minimize risk. It is sometimes also extended to maximizing opportunities. The way it is conventionally presented is very limited, for two reasons:

- In a disruptive project, an entrepreneurial endeavor always requires a certain amount of risk, so the intention to totally remove risk or minimize risk will necessarily impact project performance.
- In reality, beyond avoiding catastrophic risks, maximizing opportunities is probably a much more effective lever to success than focusing too much on risk "management."

We are not advocating for risks to be ignored, left unidentified, or neither assessed nor managed, and we are not suggesting that the project leader should not spend any time understanding and managing his project's risks. Rather, we advocate that the successful project leader must view the entire process from a different perspective. Using a different mindset, he can focus his energy on maximizing the opportunities of the project, while at the same time preventing those risks which are possible to prevent and making the best out of the intrinsic risks of the project itself. (Some risks must be taken anyway in order to achieve the project's purpose.)

Projects, being one-off endeavors, require a different risk management approach than the approach that is based on the aggregation of data from many different situations, taking place over a long period of time, like the risk management practiced by insurance firms or banks regarding financial markets. The probability of an event might be calculable with the use of some statistics, but whether an event actually happens or not during a particular project can make it or break it.

So, how does the successful project leader proceed in the field of opportunity and risk management? He uses two fundamental practices:

- He thinks of opportunity before risk;

- He understands that a small action taken today may prevent a large issue later, or may foster a great opportunity later on.

The identification process should always begin by identifying the opportunities before it proceeds to identify the risks. We are all psychologically geared towards a prudent approach, as our ancestors were focused minimizing risk to avoid life-threatening risks in the savanna. If we start by brainstorming risks, we will be swamped by the negativity arising from all the issues that could go wrong, and it will be difficult for us to revert to a positive view regarding opportunities. When we start by focusing on opportunities instead, we develop a much more balanced view of the situation. Opportunities and risk identification processes should thus always start with opportunities.

Then, the process should spend as much time on expanding the opportunities as it does on preventing risks. Organizations and projects often do not spend enough time identifying ways to leverage their opportunities, thus missing significant upsides, and sharply limiting the potential of the organization.

Following the identification of the opportunities and risks, the successful project leader knows that he needs to plant some seeds now, as early in the project as possible, to let the opportunities grow and to prevent risks over the long term. Small support-minded or prevention-minded actions can lead to a juicy harvest when the situation ripens up. The successful project leader does not hesitate to invest effort and focus on planting and supporting the growth of those seeds today. This means that the successful project leader spends enough time and attention during the project setup and preparation phases to set up the correct structures—ones that will bear fruit during the project execution. This specifically covers the project network as well as the team composition and spirit.

The choice made by the Oldavians' team with regard to the shape of the shaft is an excellent case study of an early investment done to mitigate later risk (successfully, in that case). It was not an obvious decision, and it led to additional upfront costs. Yet it was still relatively inexpensive when the decision was made, compared to the potential risk. Stakeholders were hostile to that decision, and some found it plainly ridiculous. Yet it worked!

Persistently communicating a reasoned long-term sense of optimism

The successful project leader, as an Entrepreneur, also practices another key skill: **Constantly communicating a positive vision of the long-term future.**

Once again, this does not mean that the project leader is just naively avoiding any consideration of the risks and obstacles faced by the project, and does not imply that the project leader lives in any sort of dreamland. The project leader recognizes the fact that at any moment, project team members might suffer from a particularly difficult situation. But this key skill means that the successful project leader is able to view the present difficulties as necessary evils, and is able to share the long-term positive vision of what the project will achieve. It is a very important role for the project leader, because he is often one of the only persons who can afford to have a long-term vision, one extending beyond hassles and events of day-to-day life in the office. If the project leader did not keep his eyes on the positive final outcome, and persist in actively communicating his vision to his project team, the entire team might soon be swamped by poor morale and feelings of negativity.

During our consulting assignments, we have encountered situations where the project leader—or even the top management of the company—was visibly negative and pessimistic about the outcome. This is not a useful or sustainable situation, and failure is assured as the project team, hampered by low morale, unintentionally allows the self-fulfilling prophecy to be realized. A communicative positive outlook that also stays realistic is a key practice of the successful project leader. And it is not a hypocritical position, either: positive things always come out of the deep, intimate and exciting bonds of teamwork, even if those positive things are ultimately not in the same shape or form as what the team expected at the beginning of the project.

This is quite obvious in the fable as the going gets tougher towards the end. Simon's positive outlook contrasts starkly with Harry's final negative spiral, and this, added to the Oldavians' team cohesion compared to the ultimate disbanding of the Norlanders' team, made all the difference.

Another undesirable form of leadership failure is when the project leader is stuck in reactive mode, reacting to events, and often even becoming overwhelmed by them, without being proactive. It is undesirable because the rest of the project team, and increasingly the other stakeholders, will come to feel that they have lost their helmsman. It is less obvious but much more frequent than outright negativity from the project leadership, and it can be just as destructive on the long run. If there is one person that needs to be outside the reactive mode in a project, and who needs to be proactive and build toward the final outcome, that person needs to be the project leader.

The successful project leader's entrepreneurial, long-term, win-win approach to negotiations

With his Entrepreneur mindset, the successful project leader always views negotiations with external parties (clients, suppliers, etc.) with a long-term, positive view. He also seeks a win-win situation, which is better than a one-sided compromise.

Often, and in particular in the construction and contracting industries, project leaders approach negotiations as though the end game is to take advantage of the other party. This is not a sustainable proposition and will eventually lead to some form of retaliation. It might not happen during the current project, but will happen down the line.

The successful project leader always keeps a long-term view, one which can reach even beyond the current project, in the spirit of looking for repeat business, because he knows that each professional arena is small enough to allow one meet the same people several times over one's career.

And so he approaches negotiations seeking a win-win solution. People or organizations will always go the extra mile if they are in a positive mindset regarding the project. And if the project is really challenging, it will require a lot of people to go the extra mile! This includes, of course, suppliers, clients, and other stakeholders as well.

The ultimate Entrepreneur: giving for free

The most entrepreneurial practice is to give without having a specific return or immediate interest in mind. Beyond the specific practices related to project management, the successful leader indulges in the key practice of giving. It is often easy to give away things that are of very small value to one's own self, but of tremendous value to others. Giving without a view toward short-term compensation is the most effective method, because it is the one that provides the most returns, although it might take some time to provide them.

Giving the key to building a relationship and thus the key to building the project network.

The most entrepreneurial project leader will freely give out what he has in terms of his knowledge, connections or other non-physical resources, with a long-term view toward possible returns that might benefit the project. In the first phases of the project, in particular, it is important for the project leader to invest in this giving practice.

Summary

The Entrepreneur role is about being able to invest early in order to reap larger benefits later. While it entails a certain amount of uncertainty, this practice allows the successful project leader to reap the fruits of appropriate early decisions. It is essential for the project's ultimate success, yet unfamiliar to the cost cutting cultures of most organizations.

Fourth Role: the TEAM COACH, Unleashing Team Potential

The successful project leader is like the familiar figure of the collective sports team coach. His leadership approach can potentially transform a team from a mere collection of individuals into an unstoppable force. He can lead the team to produce results that could not have been achieved without the special teamwork-centered approach and the specific practices he implements, leveraging the capabilities of each team member into an incredible collective capability.

The magical aspect of the successful Team Coach is only the persistent, disciplined application of a limited number of key practices. Done well, these practices can unleash incredible potential.

Team diversity and ego management

The first key activity of the project leader is to form his team. Sometimes organizations tend to impose specific team members on a project; still, it is vital that the project leader have a decisive word in the choice of his core team and of his team members overall. This is generally a natural consequence of the accountability of the project leader.

When forming the team, diversity is a key criterion. It is difficult because we all naturally tend to feel more comfortable with people with backgrounds and experiences similar to our own. While that helps with connecting in the short term, it is an obstacle to creativity and "thinking outside the box," as well as to creative conflict—activities that challenging projects require. The successful project leader is thus extremely attentive to ensuring a sufficient diversity in backgrounds and individual preferences.

Other characteristics are fundamental when selecting project team members. They must be hands-on individuals with a down-to-earth view of things; they must have shown persistence and initiative in solving problems in their previous experience (professional or personal); and they must be ready to invest their energy and time without limit for the benefit of the project. On that note, people who have had difficult times previously in their lives but who have still managed to be successful are generally a good bet when choosing project team members. Looking for successful professionals from underprivileged minorities can be a good approach, particularly in developing countries.

This is clearly the approach followed by Sally, Simon's right hand. She hires a team of 'renegades' recruited more for their drive than their reputation as experts. Harry, however, assembles a team of experts and recognized professionals that ultimately never really work well as a single team.

An even more important criterion is to ensure that the team members can check their ego at the door. A successful team is never a collection of divas. Members of successful teams must be sufficiently humble to be able to listen to the other team members, and capable of bearing conflict-laden discussions, while still making their contribution to the project. This is proven again and again in collective sports, where teams consisting of divas often lose against tight, cohesive teams made of less well-known players. And moreover, divas in a tightly connected team can be highly disruptive. They must be avoided. Project leadership is a collective sport.

Creating emotional connections

The first key practice of the successful project leader is to create an emotional connection between the project team members.

There are several ways to help this happen, so let us mention the three practices that experience has shown to be the most effective:

- **Unify the team around a compelling goal**, one to which team members can feel an emotional attachment over and above each individual's own goals. Share and broadly communicate the fact that the project objective is a particularly challenging one, and convince the team of this. The project team will feel connected around this unique challenge. An issue here is not to fall in the trap of placing the team in opposition to the external world, but rather to positively emphasize the objective of reaching this challenging goal.

- **Once the project goal is identified and perceived as quite challenging, let the team define its own way of dealing with it**. Let the team create, share and own the vision of what it intends to achieve. Let the team own the project and feel accountable as a team for its successful achievement.

- **Ensure that individuals connect quickly by creating situations where they have to display a measure of vulnerability**. Many teambuilding and networking activities, or icebreakers, are based on this principle. A good way is just to let people tell the story of their life, or at least part of it (it is astonishing how little we often know about our colleague's lives). This often creates tremendous respect among the individuals who form the project team,

Creating and maintaining this emotional connection is key. Who could let one teammate struggle alone if the two share an emotional connection? The successful project leader understands the importance of this factor, and is ready to spend significant time and resources to raise the level of emotional connections within the team. It is a significant investment at the beginning of a project, but it is worth every minute and every penny.

The effective team—a natural filter for effective team members

Once the project team has been formed and has started to develop a dynamic of its own, interesting things will happen. Almost systematically, particularly when the team has been composed from scratch with individuals that did not know each other well before, some people will get marginalized. And the more challenging and engaging the project goal is, the quicker it will happen. It might be that these people are not willing to engage or to participate in the overall team connection. Or it might be that they do not contribute at the level expected of them in terms of quality and timeliness and thus, they quickly lose the respect of the team. There are also plenty of other reasons why these individuals will not gel with the rest of the team.

Like Simon in the fable, the successful project leader knows that this process will happen, and that some individuals will quickly become marginalized . He will not let this situation last too long at the risk of letting it rot and then spread the rot to the rest of the team. Once it has been ascertained that an individual has been de facto rejected by the social dynamics of the project, the project leader will act. He will deal with the situation, even if the marginalized person was supposed to be a major contributor or someone difficult to move for a number of reasons external to the project.

Taking action does not necessarily mean removing this person from the team. It might mean recognizing that the person should have a different role, further from the project core team dynamics, such as the role of an expert. In these cases, though, it is often difficult to manage the person's self esteem, and the best solution is often to find an amicable and respectful way to part.

Depending on the social structure which already existed before the project started, anywhere from 10 to 30% of the project team might get marginalized, with different stages of rejection, some of them sustainable, and some not. The lowest percentages will occur in pre-existing organizations where people already know each other; a filter will be in place at the team formation stage and people will already know each other's strengths and weaknesses. Startups and projects starting from scratch or will be in the disadvantageous situation of going through a steeper team formation curve, with a higher rate of rejection. It is important in these cases that the leader be proactive in managing the situation once he has recognized that the team dynamics effectively marginalize some members.

The effective team, where roles change naturally

A well-known characteristic of effective teams is that people effectively fulfill roles which are significantly distinct from, or at least complementary to, the initial organizational setup. While it is important to delineate the roles and responsibilities at the team formation stage, the team roles will gradually evolve. Team members in effective teams will take on activities that correspond to their strengths and compensate for the weaknesses of others. Team members will even take on new roles that had not been considered in the original project organization. Team members' roles will evolve over time, eventually making the initial job descriptions, roles, and responsibilities obsolete. This will allow the team leader to fill in gaps such as soft skills, the organization of team development activities, and other temporary or permanent roles which had not been identified initially but which later appear to be vital for the project's success.

The successful project leader recognizes that roles will evolve and will not force the team to stick strictly to the originally assigned roles and responsibilities. This statement goes against the principles of management— and that is part of what makes leadership different.

Because of his theoretical approach to project management, Hugh, the Norlanders' project management expert, imposes a very rigid project organization with fixed job descriptions that people are not supposed to deviate from. On the other hand, Simon recognizes that people will naturally fall into their place on the team, and will sometimes evolve into positions that were not expected or that responded to their natural strengths. There are multiple examples in the fable: Sandra, who becomes the work site manager, the geologist who is placed in charge of the mussels issue, etc.

The new roles do not need to be written down, because an effective team knows what the other team members are up to. Responsibilities for specific actions need only to be clearly identified.

Measuring the team's pulse

Successful project leaders like Simon know how to take the team's pulse, and then take action on it.

They know how to get informal feedback from the team by walking around and discussing matters with a cross section of the team.

They also know how to get more formal feedback in the form of anonymous surveys, which allow them to know and understand which parts of the project team environment need to be taken care of. These surveys, if well structured, will allow the project leader to identify specific issues for certain categories of team members or for certain departments / specialties.

Not only does the successful project leader measure the pulse of the team, but he also takes action based on what he finds. Not on every issue, of course, because not every issue can be easily solved, but on the most problematic issues. It shows his involvement in getting the team to work together effectively, and it resolves the main hurdles that are in the way of the development of an effective team.

Maintaining an effective team discipline, Part 1: fostering creative conflict in a delivery framework

The most crucial part of team coaching is to engage the project team in an effective team discipline, a terminology coined by Patrick Lencioni. This means a set of rules and practices that ensure that the team is at its most effective delivery level.

'Effective' means demonstrating effectiveness in solving the particular problem of the project. It does not necessarily mean being the most streamlined or being as efficient as possible. Effectiveness, in the context of a challenging project, implies the ability to have dynamic discussions where options are debated and created. It implies creative conflict, which involves incredibly productive discussions that engage all participants in the confrontation of new ideas—the ultimate source of creativity.

The successful project leader is an expert in the practice of fostering and maintaining creative conflict. There are definite tools to help achieve creative conflict, like brainstorming and similar approaches. Creative conflict will happen amongst a well-connected team that is in a safe environment. In all these instances created by the project leader, his behavior is key to allowing the team to freely access its creativity. Once he has set up the proper environment, the project leader should take a back seat during the discussion, just making sure that the team stays focused on resolving the particular problem that is the focus of the project, and otherwise letting the team debate the solutions, even unconventional ones.

Around the core practice of creative conflict, there must be a delivery framework. As soon as the team has settled on a tentative solution, the wheels of execution need to start cranking immediately. And the results of action need to be fed back into the decision creation process.

Harry does not accept any challenge to his experience or his authority, or to the top-notch scientific approach to project management embodied by Hugh. He never lets creative conflict happen, even with his core team. As a result, he deprives himself of potentially useful input, as when he does not want to listen to Hannah and Hilary's pledge to simplify the follow-up system in order to improve the progress of the project. On the other hand, and although it is sometimes painful for him, Simon lets creative conflict unfold, and supports collective decision-making in the most arduous matters. He knows that what is most important is for the team to buy into the decision, and he knows that creative conflict is the best way to find unexpected solutions. For example, the idea of digging deeper to avoid boulders is suggested unexpectedly by an engineering graduate.

This can only happen if the project team applies a strict policy where, once an option is agreed upon after a heated debate, that option is unconditionally accepted and implemented by the entire team. Here, the results are measured to assess their results and to implement the necessary feedback loops. In these situations, success or failure is taken up by the team as a collective.

Some project teams, particularly large ones, like to write down the expected principles of behavior in a project charter that can be posted in the project office and shared with newcomers. Like any values posted on the wall, this behavior charter is only as good as the leader's and core team's behavior; it does not relieve the team from its responsibility in terms of setting the correct example.

Maintaining an effective team discipline, Part 2: delivering on commitments

After actions have been defined through a creative conflict process, written down, and adopted by the team, the project leader needs to maintain the discipline of follow-through.

This requires two practices.

The first one is to limit the number of actions to a manageable number. Actions therefore need to be focused on those high impact actions that have a deep effect on the project delivery, namely, actions that tackle the project constraints.

The second one is to relentlessly check that the actions are both effectively taken and then followed up. This implies both being disciplined regarding the criteria for close-out, and looking for regular candid updates.

Too often, actions are considered closed when they are 90% complete. This is not acceptable in a project; the last 10% of progress is often extremely painful. The successful project leader only accepts 100% completion.

Additionally, action close-out needs to be demonstrable by tangible evidence. This is why it should generally be either a clear physical deliverable or a measurable outcome.

Keeping the momentum when it gets tough

The cohesiveness of a team is often measured by how cohesive it stays in difficult moments. All teams feel good when everything goes well. In tough times, weak project teams will fall into a culture of blame, split into competing camps, and ultimately, everybody will try to avoid responsibility for whatever caused the difficulty or even for the project's failure. Strong project teams will not fall apart during these moments, but rather will assume the responsibility for these issues collectively; they will not start blaming each other, or the external world, for what is happening.

The successful project leader will make sure to redouble his team cohesion efforts in these instances, and will not allow himself to fall into the blame trap. Maintaining the team emotional connection and enhancing the positive successes, however small, are also key to the practice of team coaching.

It is when things get really tough that the differences between the weak and the strong team will be seen; and it is at these times—which will necessarily happen during a challenging project—that the investment put into the good preparatory work will start to bear fruit.

In the fable, Simon invests heavily in the cohesion of the team during the team formation stage, and invests further during the execution of the project. He manages to have fruitful brainstorming sessions on the most difficult issues without pointing fingers. He reaps the benefit from all the investment he put into the team at the decisive moment at the end of the contest. Harry, on the other hand, spends his time finger-pointing, and the result is an ineffective working atmosphere to the point of causing a mutiny at the construction site, and an overall disbanding of the team when it becomes clear that success is no longer an option.

Be an effective barrier to the stakeholder-induced stress

While a significant part of project delivery involves keeping the stakeholders happy, a vital part of the role of the successful project leader is to provide an effective barrier protecting the project team from haphazard stakeholder requests.

Nothing is worse than a team leader giving way to any request from stakeholders and then passing on the stress to the rest of the team. The focus of the team will then move away from the actual project delivery, which will ultimately end with a poor final result.

Like Simon, the successful project leader knows how to take a stand when needed to protect the team, while at the same time, knows to protect the project's interests by doing what is necessary for the key stakeholders. This is a fine line that requires discipline and a strong sense of self-awareness, in particular to avoid transmitting one's stress to the project team.

Being able to have tough conversations with individuals and with the team

It can get tough for the project team. It can also get tough inside the project when the project leader needs to have tough conversations with individual team members or with groups. This might be because of their performance or because of the social dynamics of the group, but no matter what the reason, these conversations can have a great impact on performance.

This is an area where a significant distinction exists between the average project manager and the successful project leader. Most managers and leaders dread those tough conversations to the point of avoiding them completely. The successful project leader, however, has no fear of these tough conversations. He uses discretion, waiting until the conditions are right for a conversation, and he will set up the right context so as to make these conversations as effective as possible. Still, he does not wait so long that the situation festers or the conversation becomes impossible. Because of ego issues, tough conversations don't work with divas, or only work after a significant and dramatic failure, which should be avoided if at all possible.

These tough conversations, in the right context, lead to an effective change of behavior and performance and are definitely differentiate the successful coach from the weak project leader. Choosing the right mix of team members, managing the team formation and dynamics, and ensuring team discipline, are necessary ingredients, but they are not sufficient. Having tough, candid conversations when they are needed is the final ingredient that creates an effective team and that shapes the effective Team Coach.

Summary

The Team Coach role is about creating an effective team, a team that will be successful and keep momentum and cohesion when facing long odds. The successful project leader knows how to create a diverse team and then foster its development by allowing roles to evolve and by letting people who don't fit leave the team amicably. The project leader also knows when and how to have tough conversations with team members when this becomes necessary.

Fifth Role: the PEOPLE CATALYST, Releasing Each Individual's Talents

The successful project leader is not only an outstanding Team Coach, he also knows how to uncover people's individual talents, and knows to support the expression of each individual's potential.

His appreciative way of looking at each individual's talents will sometimes be destabilizing, because he will look beyond the conventional professional identity of people to appreciate their overall potential. To the remote observer, strange things will happen as people suddenly start doing activities that are quite distinct from what they would have been expected to do as professionals in their narrow field.

Through this appreciative action, the project leader not only fosters talents that will ensure the success of the project, but he will also create a strong emotional bond with the person which will last far beyond the completion of the project. This puts the seed of people's future development in a radically entrepreneurial act.

The People Catalyst role is different from the Team Coach role in that it does not address how to multiply the intrinsic set of talents in the team through team practices, but rather discusses how to develop the individual talent of team members.

Knowing the people

The successful project leader knows the people in his team. Although it seems a straightforward statement, experience shows that overall, managers in the conventional sense do not know a lot about the people in their team; moved by a somewhat mechanistic view of work, they do not take the time to look beneath the outer appearance of their employees.

Successful project leaders take the time to discover the actual people below the surface, and are able to develop a true emotional connection on a more personal level as a result. This is crucial, because without this relationship quality, without this initial connection at an intimate, emotional level, the project leader cannot envisage any of the actions that are discussed in this section.

Personal relationships can be developed outside work or during activities of common interest. Pictures of a family on the desk can also provide opportunities for more personal discussion. The successful project leader knows how to take the time to be fully present for his team members, to listen to them, and to give them the feeling that they are special. These discussions do not need to be long—everybody understands that the project leader is busy—they just need to be intensely focused.

Taking stock of people's potential: the appreciative approach

The successful project leader takes an appreciative approach when dealing with the people he associates with.

In conventional organizations, it is commonplace to develop a rather negative view of people—highlighting those areas where they are not so good or even areas where they are frankly bad. People also tend to focus on problem areas instead of on areas that are good or at least improving.

The 'appreciative inquiry' approach is a relatively new approach that was developed in the psychology field in the 1970s. It is closely related to 'positive psychology,' which developed around the same time and became mainstream in the 1990s. It is based on an approach that seeks to uncover what is working, to discover what is good, and to show appreciation for these positive aspects and develop actions based on these positive observations. By having this different mindset of focusing on what is working—and doing more of it, instead of focusing on trying to solve problems or overcome weaknesses—appreciative inquiry is a really powerful tool.

Not everybody is suited for any particular kind of activity or position. It is important to make sure the role allows the person to express his talents and make a difference. The effect is twofold: the contribution to the project is more intense, and the motivation and energy of the person is multiplied, multiplying his effectiveness.

It often happens that the position that a person applies for, which is his trade, does not really reflect the true extent of his talents. From my coaching experience, I would say that in general, conventional organizations only use 20 to 30% of their true potential with regard to people's talents, and disdain the rest. Why is this? It is primarily an effect of the entire Industrial Age education and employment system, which has been reinforced by the bureaucracy of many large organizations. People need to fit into career paths and trades, and need to be neatly categorized in boxes designed by organizational gurus. As a result, the organization ignores their other talents, even when these talents could prove extremely useful.

Now, not all people's talents are usable for any given project. Still, many can be used for one occasion or another, as long as the talents are known. The project secretary's talent of singing soprano will not directly influence the project execution, but might create an excellent source of emotional linkage at a teambuilding exercise or at an ice-breaker with a client. A project engineer's passion for supporting a humanitarian cause will not serve the project directly, but could create a better team dynamic or provide the team with connections to useful other people and to new ways of working in a particularly remote country.

Imagine if, instead of using only 30% of people's talents on average, we could use 50-60%! That would definitely change the game. This is what successful project leaders seek to achieve, thereby bringing extraordinary value to the organization.

Develop your strengths; don't try to correct your weaknesses

This statement goes deeply against the long-held wisdom that correcting weaknesses should be the main objective. In the Industrial Age, that approach of focusing on correcting weaknesses first made perfect sense, as all people were supposed to fit some pre-determined bill, and needed the right dimensions to work as a cog in the overall machine.

Today in general, and even more in a well-functioning team, correcting weaknesses should not be the main priority. Others will compensate for your weaknesses, as long as is the weakness does not make communication a total impossibilityand as long as the weakness is not a deep-seated flaw that makes it impossible to be a functioning part of the team. Since one's energy is limited, why not spend it on developing your strengths instead of slightly reducing your weaknesses?

This is a radical idea, and proves extremely effective when it is really put into practice. The successful project leader follows that path and seeks to develop team members' strengths in order to achieve world-class contributions, but he also lets team members' weaknesses be compensated for spontaneously by the strengths of other team members—strengths which are available thanks to the diversity of the team.

The art of deep one-on-one conversations

In the Team Coach role, we have touched on the fact that being able to have tough conversations involving candid feedback is a key practice for the successful project leader.

One could believe that having appreciative and supportive one-on-one conversations is emotionally easier. But genuinely deep and appreciative conversations can be really difficult to achieve when they involve one's personal identity; and this, too, is a key practice of the successful project leader. As with one-to-one coaching sessions, these conversations must involve a lot of active listening, a deep personal connection and a deep sense for inquiry.

It is far from easy to identify the real talents and sense of purpose of an individual, and it is even less easy to have the individual accept the new identity that is revealed. Society, and/or family, has often formed us in such a way that our identity is closely linked to our professional identity. (Think, for example, about how you present yourself to a stranger.) During this deep investigation, it might be that you will uncover that your true identity is quite different, or even radically different, from the one you generally imagine for yourself. For example, perhaps you thought of yourself as an accountant and defined yourself as an extremely professional and serious accountant, and then suddenly you realize that you want to be on stage, doing public speaking. Your coworkers support your efforts to experiment with this new sensation, and you find public speaking exciting. You receive the feedback that you are good at doing it, and you connect emotionally with the audience. Suddenly, you're struck by the discrepancy between your normal identity and this new field of talent. Will you go through the big step of changing your identity?

In the fable, Simon manages to have a series of close, positive conversation with Sandra that allows her to overcome her self-limiting beliefs about becoming a work site leader. Not every manager can do that. It takes real leadership and self-awareness, and the capability to recognize and appreciate the potential of the people you work with.

The successful project leader will not be afraid to have the conversations that will allow you to bring your full talent to the team by relieving you of some of the constraints you place on yourself or that society or family have placed upon you. It might be that you will ultimately choose to stay as you were in your previous role, or it might be that you will change your life by broadening your horizons. The successful project leader has the ability and the practice to conduct these potentially life-changing conversations. These conversations take place at the deepest level by bringing you to question your identity. Having these conversations in a safe manner is difficult and uncommon in the workplace, but because of the benefits that come from knowing what talents are available, this is a skill and a valued practice of the successful project leader.

Supporting the expression of people's talent, in service of the project team

The successful project leader is thus able to consider and appreciate each individual's talents and to figure out how they can best contribute to the project. This approach is based on two fundamental principles of Soft Power leadership:

- People will be much more engaged and effective if they do what they are passionate about.
- In a large and diverse team, there is always someone who knows the solution to a given problem.

The proper role of the successful project leader is thus to engineer the project team into a place where people can express their full potential. Compared to a conventional manufacturing organization, it is much easier to do that in a project which is fluid by its nature, and where opportunities appear all the time to take responsibilities and act on unforeseen events. Knowing the people on a project team in depth is key to accessing the right talents and resources whenever they are needed.

In the fable, Simon benefits several times from his team's unexpected skills and knowledge. For example, the geologist with a Masters in Marine Biology is extremely useful during the mussels event. Because team members are considered to be filling strict roles on the Norlanders' team, this never happens under Harry's leadership.

This also means that the project leader leads people to take measured risks, while ensuring that these risks will not be fatal to the project or the person and also supporting people even if they are not successful the first time they try to use their talents. It is a measure of challenging people enough to let them grow, but not so much that the difficulty becomes unbearable. It is about supporting people so they can reveal their talents and discover more about themselves. In the fable, Simon takes the risk of pushing Sandra to greater heights based on her talents, with some difficulty along the way, and eventually being rewarded with great results.

Helping individuals reveal their talents and their potential is generally a long-term endeavor. It is an entrepreneurial venture that involves helping people to overcome their psychological obstacles (such as self-limiting beliefs and other blind spots) and to improve their self-awareness, while propelling them toward situations where they can effectively exercise their talents in the service of the project.

The synergic relationship between Team Coach and People Catalyst

The function of the project leader as both Team Coach and as People Catalyst is deeply intertwined. Still, we have chosen to present them as two different dimensions of the practices followed by successful project leader, because the visible practices they entail are quite different. Team coaching is a highly visible position, as it involves unleashing the collective genius of the team and making sure the group transcends the contributions of individuals. The People Catalyst starts from the other end by identifying and promoting latent talents in individuals and making them contribute in unexpected ways to the collective project endeavor.

Starting from opposite sides—the team and the individual—those two roles become even more powerful when they synergistically meet and create a chain reaction of astoundingly successful projects.

Today, personal development is a significant component of compensation

Giving development opportunities to people are not only a requisite to attract good people; it can become an attraction point in and of itself if the successful project leader has a known track record of developing and transforming his team members.

In a down-to-earth consideration, people today expect personal development to be a significant part of the package when they accept a new job. In a way, personal development is now part and parcel of the compensation package, and can often be used to partially offset monetary compensation. On the other hand, talented people with high potential will not take positions where no development opportunities are provided, even with an excellent compensation package. They will be more attracted by positions where they will be challenged and be able to significantly develop their abilities and their track record.

The conclusion is obvious: if you want to attract great people, personal development needs to be a visible element of the package. The successful project leader makes that part of the package credible and follows up.

Conclusion

The People Catalyst role is a key role in attracting, retaining and uncovering the latent talent of worthwhile people. By applying this role consistently and generously, the successful project leader creates other future successful leaders, and at the same time, reaps the benefits of having discovered their nascent new talents.

Beyond their contribution to a particular project, he plants the seed of a thriving and vibrant community of successful project and organization leaders. Should the organization recognize it, and take into account the unconventional paths followed by the most successful elements of the project team, the organization can then count on a pipeline of talented personalities that will deliver incredible results in the future.

The People Catalyst role is the key to present success and future performance.

Summary

The People Catalyst role is about developing a deep and appreciative view of people and identifying how their talents could support the project. Beyond people's professional identities, the successful project leader can catalyze revolutionary growth in his team member's roles and identities, bringing them and the project tremendous value.

Synthesis: the Holistic Project Soft Power Practitioner

The symbiosis of the five roles

The two roles of the Entrepreneur and the Kung-Fu Master are mainly about personal mastery.

The two roles of the Team Coach and of the People Catalyst are mainly about creating and maintaining deep emotional connections with individuals on the project team, creating extraordinary value.

The first role, the role of the Spider weaving its network, is the central role of the successful project leader, and contains both elements of personal emotional work, and collective emotional work—in the sense of developing deep connections with other people.

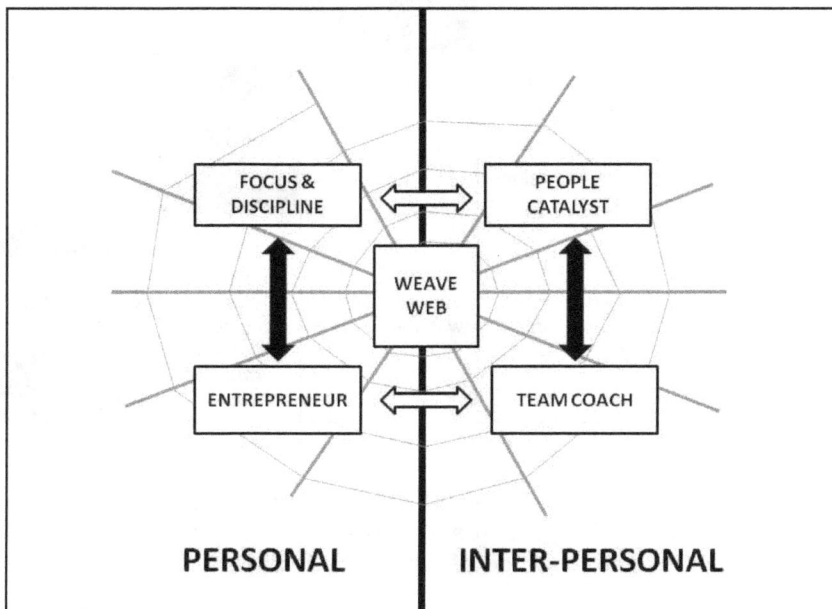

Studied carefully, the five key roles of the successful project leader appear to be deeply intertwined. Each one of them is deeply informed by and related to the others. This is best expressed in the following table.

KUNG FU MASTER				
Longer term focus, values / Alignment of actions with goal	ENTRE-PRENEUR			
Prioritization of stakeholders	Longer term focus in all network relationships	NETWORK WEAVING SPIDER		
Focus on results / challenges	Get superior value in the long term by investing in team formation stage	Weave network with the team	TEAM COACH	
Focus on developing people talents for the project: focus on the strengths not the weaknesses	Invest in people for the longer term	Weave network with the individuals	Find out the talents to contribute out of their normal remit	PEOPLE CATALYST

Still, it is only the practice of the five roles concurrently that creates the rounded project leader, and exemplifies the practice of Project Soft Power.

Influencing the future in a positive way

The practices of the successful project leader are actually founded in one major philosophical belief which is at the basis of all action. This belief can be summarized thus: *It is possible to influence future outcome in a positive way.* There is also a corollary: *People have a huge, undisclosed potential.* From this main principle comes the importance of defining the project purpose and objectives). From the implementation of this belief, there arose a number of practices involving assertiveness, personal discipline and postponed gratification to reach the expected objectives. As a consequence of the corollary, the positive view of people's potential expresses itself in practices that revolve around deep interest in people and around team work enhancement, so as to allow people's potential to be expressed to the world for the sake of the project.

The key personal practices can be learned

One important observation is that these key personal practices are not innate talents. Of course, we might be more or less talented in their practice, due to our education or our past experiences. Still, anybody can learn, practice and improve these practices to the point of mastery. Some of these practices may be more difficult to master than others for one particular individual, but still, with the right amount of effort and time, mastery could be achieved.

It might not be effective to develop all of these five practices up to the point of mastery, though. It might be much more effective to develop the one or two practices which we feel naturally more comfortable with. As for the others, we need to understand them deeply, and make sure that we are not so weak in them that they become a liability. But then, instead of spending a tremendous amount of energy to elevate our skills in all five areas to mastery, it will be more effective to compensate for our areas of relative weakness by using the skills and talents of other people in the team.

The next steps

Everybody can become a master of Project Soft Power, developing some skills to mastery and knowing how to compensate for others. It takes effort and time to achieve that goal, as with all personal improvement programs. I know that although you might concur with the ideas presented in the book so far, at this stage there is only a small probability that you'll implement the disciplines consistently over time and become a Project Soft Power master. The rest of the book is devoted to ways which you can use to increase that probability and that give you the directions and the tools to become a Project Soft Power master more quickly.

First, let us respond to the question: How do you fare in terms of Project Soft Power? In Section 3, the Project Soft Power self-test will allow you to measure your familiarity with the five roles of Project Soft Power.

Section 4 will then act much like a personal coach, transferring these findings into actions and making recommendations.

Finally, in Section 5 of the book, we will go into deeper detail regarding the emotional issues that the successful project leader must overcome when practicing Project Soft Power. And from there, you will identify which is the exercise regime that you need to follow to significantly improve your Project Soft Power profile.

Useful References for Project Soft Power

So as to not overwhelm the reader, here is only a short list of select references that are highly recommended to those who wish to further study the issues at stake in Soft Power Leadership.

Network building and effectiveness

The Tipping Point: How Little Things Can Make a Big Difference, Malcolm Gladwell, 2002

The Wisdom of Crowds, James Surowiecki, 2006

Focus and discipline

The Goal: A Process of Ongoing Improvement, Eliyahu Goldratt, Jeff Cox, 1984

Theory of Constraints, Eliyahu Goldratt, 1999

Critical Chain Project Management, 2nd ed., Lawrence P. Leach, 2005

Be Fast or Be Gone, Racing the Clock with Critical Chain Project Management, Andreas Scherer, 2011

Project Management in the Fast Lane: Applying the theory of constraints, Robert C. Newbold, 1998

Great by Choice: Uncertainty, Chaos and Luck – Why Some Thrive Despite Them All, Jim Collins, 2011

The 80/20 Principle: the Secret of Achieving More with Less, Richard Koch, 1999

Entrepreneurship

Mavericks at Work: Why the Most Original Minds in Business Win, William C. Taylor & Polly Labarre, 2006

Team effectiveness

The Five Dysfunctions of a Team: A Leadership Fable, Patrick Lencioni, 2002

Overcoming the Five Dysfunctions of a Team: A Field Guide for Leaders, Managers and Facilitators, Patrick Lencioni, 2005

Built to Last, the Successful Habits of Visionary Companies, Jim Collins & Jerry Porras, 1997

The Wisdom of Teams: Creating the High Performance Organization, Jon Katzenbach & Douglas Smith, latest edition 2003

People catalyst

Appreciative Intelligence: Seeing the Mighty Oak in the Acorn, Tojo Thatchenkery & Carol Metzker, 2006

Project Soft Power Self-Assessment Test

Introduction

The following simple self-assessment test will allow you to clearly identify how familiar you currently are with each of the five different roles of Project Soft Power.

A more convenient, printable version it is available for free on the Internet at the website www.ProjectSoftPower.com (we're also working on an online test for the future), along with the Project Soft Power workbook for devising your personal Project Soft Power action plan.

You should already have an intuitive feeling of which roles are more familiar to you; the test will simply serve to add a practical measurement to your intuition. This self-assessment is there to allow you to know yourself better; hence candor is required. It is preferable to answer each question relatively quickly, as your first views often better reflect your unfiltered preferences.

As an advanced version, in addition to taking the test by yourself, ask someone who knows you very well, like your significant other, to perform the test as they see you, and compare the result. Discrepancies in the result will give you significant insights about potential blind spots, those areas which you might not realize are lacking or excessive.

In another advanced application, the test can also be run as a 360 degrees assessment; as implementation and interpretation of the results will require professional help, please contact us should you wish to perform this version of the test on project leaders or within a project core team.

The Test

The following pages contain the Project Soft Power simplified test. The test is designed for a project leader; however you can apply to your case by looking specifically at your own remit.

Some questions request an evaluation of parameters. For these questions, we encourage you to take a look at your project records rather than to answer off the top of your head, as reality and your perceptions might be somewhat different.

Actually, if you can, to get a more objective view, we recommend that the questionnaire be filled by someone who is working closely with you, rather than by yourself. Or at least, have some of these questions answered by close co-workers to check whether your answers are consistent with what they have observed.

Get an answer to the questions, then use the score sheet on the next page to copy your answers in the right places. Some answers will be copied at several locations as the questions simultaneously address several Project Soft Power dimensions. Add in the columns to discover your Project Soft Power score. Note the numbers associated with relative strength and weakness. The following Section of the book will explain how to use this result.

	1	2	3	4
1	**How clear is the purpose / objectives of your part of the project (or of the project if you are the project manager)?**			
	There are no clear objectives	There is a clear high level objective, no detailed objectives	Some detailed objectives have been developed but not communicated	Detailed objectives are developed and communicated
	1 ☐	2 ☐	3 ☐	4 ☐
2	**What is the percentage of your team members whose past professional and personal experiences you know in detail?**			
	Almost nobody	Only the core team	About half	Almost everybody
	1 ☐	2 ☐	3 ☐	4 ☐
3	**How often do you visit the worksite(s) / subcontractor's sites?**			
	Almost never except formal high-level visits	Rarely	Every one or two months or so	At least every fortnight
	1 ☐	2 ☐	3 ☐	4 ☐
4	**How regular are you with meetings and discussions with your core team?**			
	Very irregular, only when I have time	Irregular depending on the workload	Quite consistent except if I am travelling	I make it a point to be consistent even if I am travelling
	1 ☐	2 ☐	3 ☐	4 ☐
5	**How much time do you spend networking for your project scope with the relevant external stakeholders?**			
	Almost never	Rarely, on occasions	About 10-15% of my time	More than 30% of my time
	1 ☐	2 ☐	3 ☐	4 ☐
6	**How often do you sponsor your team members' hobbies / interests in the name of the project (in time or money)?**			
	Never	Rarely	Sometimes	Frequently
	1 ☐	2 ☐	3 ☐	4 ☐
7	**How punctual are you generally?**			
	I am known to be running always late	I am known to be quite flexible with my timing and to have regularly prolonged meetings	I am quite punctual although I occasionally let meetings go much beyond their timing	I am known to be extremely punctual and to keep meetings within their time
	1 ☐	2 ☐	3 ☐	4 ☐

	1	2	3	4
8	You have the possibility to improve the project bottom-line by a value of 100 in one year time, by investing today in a different option. However, there is a 20% risk that the bottom-line improvement will not appear. How much are you ready do invest in extra cost today?			
	I won't invest 1 ☐	I will invest if it costs less than 20 2 ☐	I will invest if it costs between 20 and 60 3 ☐	I will invest even if it costs up to 60 and more 4 ☐
9	How often do you measure your team's pulse? (using methods such as online surveys, 3rd party surveys etc)			
	Never 1 ☐	Rarely, if I see that there are some problems 2 ☐	Sometimes, when there are significant changes 3 ☐	Regularly, every two months or so 4 ☐
10	What is the proportion of identified opportunities vs identified risks for your part of the project?			
	Less than 10% 1 ☐	Between 10 and 20% 2 ☐	Between 20 and 40% 3 ☐	More than 40% 4 ☐
11	What is the percentage of people in your team that have been promoted during the project / just after the project completion?			
	None 1 ☐	Some rare individuals 2 ☐	A sizeable proportion (20-50%) 3 ☐	Most of the team (more than 50%) 4 ☐
12	What percentage of actions related to risks and opportunities have you identified as 'high priority'?			
	More than 50% 1 ☐	Between 20 and 50% 2 ☐	Between 10 and 20% 3 ☐	Less than 10% 4 ☐
13	How much time are you actually seated in your office doing your own work when you are in the project premises?			
	Almost all the time 1 ☐	Most of the time except for meetings 2 ☐	Quite rarely 3 ☐	People never manage to catch me in my office 4 ☐

	1	2	3	4
14	**How much do you invest yourself in the project setup (systems & processes) for your part of the project?**			
	Not at all	Only when there are project particularities to solve	Somewhat to make sure that things are being organized properly	I drive the setup of the systems and processes
	1 ☐	2 ☐	3 ☐	4 ☐
15	**How often do project team members take on positions or responsibilities that are significantly different or even radically at odds with what they were supposed to do?**			
	Never	Very rarely	Sometimes	Often
	1 ☐	2 ☐	3 ☐	4 ☐
16	**How much was your team involved in the definition of your team's objectives / purpose?**			
	Not at all	Consulted on a document	Team was asked to generate ideas but objectives were developed with the sponsor	Fully involved (participative workshop)
	1 ☐	2 ☐	3 ☐	4 ☐
17	**What percentage of the team is aware of the current main project priority?**			
	Almost none	The core team	About half of the team through direct communication	Everybody (public communi-cation)
	1 ☐	2 ☐	3 ☐	4 ☐
18	**What is the frequency of contacting your private advisory council?**			
	Never	Sometimes	Every month	Every week
	1 ☐	2 ☐	3 ☐	4 ☐

Project Soft Power Simplified Test Results

Following the number of the question, write your answer in each box. Then add the number in each column to discover your Project Soft Power scores.

1:	4:	1:	2:	2:
3:	7:	8:	4:	6:
5:	12:	10:	9:	11:
9:	14:	14:	13:	13:
18:	17:	16:	16:	15:
↓ Sum each column below ↓				
SPIDER	KUNG FU	ENTRE PRENEUR	TEAM COACH	PEOPLE CATALYST

The minimum score for each category is 5, and the maximum score is 20.

A score of 10 or below indicates a weak spot; a score of 16 or above indicates a strong point.

My two top strong points are:

Write here any weak points with a score of 10 or below if any:

How to Improve your Project Soft Power

"The way you deal with your own failures and shortcomings sets the tone and example for how the rest of your team and organization acts as well"

– Charlene Li, in Open leadership

View this section as your personal coach on improving your Project Soft Power. With the self-assessment test, you now know where your strengths and weaknesses lie. It is now time to reflect on these observations and commit to a course of action.

Don't be impressed by the length of this Section, you will only need to consult those few pages that are relevant to your case.

What should I do with the test results?

There are, generally, two philosophies with self-assessment test results.

The conventional approach is to identify those areas which are weaker and focus on improving them. This approach has the advantage of raising the bar to avoid really weak areas. It has a major drawback, however—it tends to bring you to a situation where you are average in all areas, where you will tend not to be distinct from your peers. It tends to commoditize your talents; and if you burn your available energy to improve weak areas, you won't shine in any.

Which is why we recommend a second, less conventional approach: identify those areas where you are particularly strong, and focus on improving them. Just be aware of those areas where you are weak, and find a way to compensate for them.

There is a caveat, however. If you have a really rock-bottom rating in one area (typically, less than 10 in the book's self-assessment), then work to raise your skill in this area to a minimum level to avoid suffering from a blind spot effect, but don't try to bring it to a higher level. That would swallow most of your energy and result in limited improvement, as weak areas are probably areas that are not your preferred way of operating. Changing personal preferences can be done, but is generally a very long-term endeavor spanning over years. This is why it is not generally a good solution for improving your personal effectiveness over the short and medium term.

Don't develop weak areas—instead compensate for them

You should compensate for those areas where you are weak. How can we compensate for weak areas? There are several ways to do that; in projects, which are collective endeavors, the easiest is to work with people who are strong in those areas where you are weak and let them act on your behalf. This principle works particularly well when setting up project core teams. It might be difficult because these people will approach problems differently and you might be a bit estranged at first because of these differences; learn to view them instead as a source of enrichment and effectiveness. It is linked to the fact that the most effective teams are generally the most diverse.

There are other ways to compensate for your weak areas, much as handicapped people often find inventive ways to overcome their handicap to perform tasks that other people would perform easily. This approach generally boils down to disciplined practice, goal setting, or even creative ways to go around the particular difficulty one encounters when performing specific tasks. For example, Richard Branson is dyslexic; it does not prevent him from giving speeches, which are an important part of his leadership practice; he gets them printed in a particular way, and probably memorizes more of it than the average speaker would.

Further developing your strengths

The following pages will give you some hints about how to further develop your Project Soft Power strengths. These are the two strengths you have identified in the Section 3 test. Don't read everything: go directly to the relevant page for your case.

How can I boost my Spider mindset and behavior?

If you have identified the Spider role as being one of your strengths, you are naturally good at networking and connecting with people.

How can you enhance your skill further? Ask yourself (or, with the help of a 3rd party acting as a coach) these simple questions:

- What makes me successful in connecting with people?
- What are some of my practices that I notice are particularly effective in connecting with people (versus others that are less effective)?
- How effective am I in connecting with the right people that effectively count for the success of the project?

Often, people who are strong Spiders will be particularly good in the connecting/personal relationship building and maintaining part. They often can further improve their effectiveness for the project by networking and connecting in a more systematic way, keeping a purpose in mind. The aim is to create a project-specific cobweb based on the influence potential of the stakeholders. Of course, that should not be stretched to the extreme where one only does the connecting effort for a reason; experience shows that connecting and building network needs to have a certain dose of serendipity. Still, by focusing one's effort and time on stakeholders that are important for the project outcome, rather than on a feel-good factor, the successful project leader does greatly improve his effectiveness as a Spider.

How can I boost my Kung Fu Master mindset and behavior?

If you have identified the Kung Fu Master role as being one of your strengths, you are intrinsically a disciplined, self-motivated person.

How can you improve your skill further? Ask yourself (or, with the help of a 3rd party acting as a coach) these simple questions:

- What makes me successful in being focused and disciplined?

- What is the process by which I decide what is important and should get my attention? How can I improve it?

- How do I feel when I decide not to follow up on a priority? What makes me avoid feeling bad about it?

- What is my method that I use to follow the discipline of consistent, repeated action and practice? How can I improve it?

- How can I improve the effectiveness of my discipline and focus for the project?

Often, a layer of strict, organized and regular calendar alarms followed by action can significantly improve one's intrinsic strength in discipline; while the "two plus one" priority practice, in public, can improve significantly the focus strength.

Often, people who are strong in self-discipline might not be so strong in terms of focus, and vice-versa. Ask yourself whether you are more disciplined or more focused.

Let's explore these two dimensions.

Very disciplined people are often also very analytical and future-oriented and are not led astray by short-term benefits. Because they are analytical, they might fall into the "analytical trap" described in the Kung Fu Master section, and not focus their efforts sharply enough on the important things that will be decisive in overcoming the project difficulties. The appropriate questions to ask yourself are these: What can I do to improve my effectiveness by being more focused? Why is it so difficult for me to let go of diving into the details?

Very focused people, on the other hand, might not be so strong on consistent discipline. Their focus-orientation is leading them to handle the important issues they have identified, sometimes sacrificing the consistent, regular discipline of project management. The question to ask yourself in this case is: What can I do to make sure that I practice project processes in an even more disciplined way?

How can I boost my Entrepreneur mindset and behavior?

If you have identified the Entrepreneur role as being one of your strengths, you are intrinsically a future-oriented person who likes to build things.

How can you improve your skill further? Ask yourself (or, with the help of a 3rd party acting as a coach) these simple questions:

- What entices me to prefer long term gratification to short term gratification? How do I feel when I build something for the long term? How do I feel when I make a short-sighted decision that is not so good for the longer term?

- How do people respond to my tendency to always think long term? What do I do to convince them that it is the right decision?

- How can I further leverage my Entrepreneur mindset for the success of the project?

Often, people who are Entrepreneurs tend to forget the need to celebrate achievements and take advantage of the present moment. They live in the future, always straining and thinking toward the next event. As they forget this important part of building the team's connection, they might become estranged from the rest of the team. Maintaining open communication channels and explaining the rationale for decisions is then vital to the success of the project leader. Ask yourself / your team the following question: How can I help the people on my team to understand the need for a long-term vision? How can I still recognize the present achievements without being deterred from the long-term goal?

How can I boost my Team Coach mindset and behavior?

If you have identified the Team Coach role as being one of your strengths, getting teams of people to work effectively together is natural to you.

How can you improve your skill further? Ask yourself (or, with the help of a 3rd party acting as a coach) these simple questions:

- What particular practice makes me successful in getting teams of people to work together effectively? How can I develop it further?

- What do people like about me that makes them happy to work in my team? How can I do more of it?

- What do people like about me that makes them happy to have me in their team? How can I do more of it?

- How does my Team Coach talent make my project effective? How can I further leverage this talent for the sake of the project delivery?

A difficulty often encountered by people who are known to be good team leaders is to have real tough conversations and maintain a strict discipline. It is relatively easy to be a popular team leader while maintaining an ineffective team. It is much tougher to evolve the team into a success machine. The successful project leader also knows how to have those difficult conversations that are needed at times with individuals, or with the whole team. He also knows that maintaining discipline and consistent application of processes is key. Ask yourself: How much is your popularity and respect based on the capability to have tough, candid conversations with your team and team members? What prevents you from delivering candid feedback? How can you further improve the team impact and effectiveness? How can you accelerate the team formation to deliver exceptional value earlier?

Another hurdle often faced by the Team Coach is the difficulty of accepting that people's roles evolve throughout the project, in particular when the team is tightly knit together. This runs contrary to the basic management principles inherited from the 'Industrial Age' (the job description). Yet this is not only a reality, it is a fundamental ingredient to team success. Ask yourself: How do I consider and recognize people changing roles to fit the project requirements? What recognition do the team members feel they are getting when they do necessary activities outside their normal scope? What can I do to accelerate this role development toward the performing team?

How can I boost my People Catalyst mindset and behavior?

If you have identified the People Catalyst as being one of your strengths, you certainly like to care about people, recognize their individual greatness and develop them on an individual basis. And people do in effect recognize it also.

How can you improve your skill further? Ask yourself (or, with the help of a 3rd party acting as a coach) these simple questions:

- What can I do to accelerate the identification and release of people's skills and abilities for the sake of the project?

- What can I do to accelerate my intimate knowledge of team members and have deep conversations with them?

- What is it in my People Catalyst skill that is most important for effective project delivery? How can I further develop this skill to make the project delivery more effective?

An issue often encountered by the People Catalyst is the difficulty of making the link between identifying and appreciating people's talents, and getting this talent to work for the benefit of the project. How people's talent will contribute to the project delivery can be very indirect (like, contributing to the team spirit only). The recognition of certain hidden talents can lead to a considerable reshuffle of the responsibilities within the team—bypassing what some believe are rights given by duration of service or by experience. Still, the recognition and publicity of this talent is key to the motivation of the team members and thus ultimately to their dedication to get the project done. The power of this motivation engine should not be underestimated. Ask yourself: How can I make this talent contribute to the effective delivery of the project? How can I use those hidden talents to significantly increase commitment and dedication to the project?

Bringing your really weak areas to a minimum level

In case you have identified yourself as being really weak in an area, here are some hints to bring your skill to the minimum acceptable level. The objective is not to become great at this skill; it is only to bring it to the minimum level so that it does not become a liability.

The following paragraphs are not intended to give you a ready-made solution, but rather to encourage you to find the solution that suits you by relying on the appreciative approach. What has worked in the past, if only once, that allowed you to reach a better level in that dimension?

The questions in the following pages are best considered with the help of a third party, such as a coach, with whom you can confidently discuss your personal issues.

Don't read everything; go directly to the relevant page for your case.

How can I show a minimum Spider mindset and behavior?

If your score is that low, you consider your networking capabilities to be minimal, and maybe you are afraid to proactively seek ways to connect with people.

First of all, let's do a reality check. Our experience suggests that people often have more networking resources available than they think. Do you have an account on a social network like LinkedIn or Facebook? Are you a member of professional associations? Are you member of one or several alumni associations? Rate your networking resources as they really are. Ask people who know you what they think of your networking skills to get a really candid assessment of your present situation.

If you confirm the observation of your weak rating, here are some powerful questions we suggest you ponder. It is much more effective to discuss them with a third party that will challenge you to respond candidly:

- What is really preventing me from developing my networking capabilities?

- In which conditions do I really freeze when contacting people? What can I do to avoid being in these conditions? [example: some people might not like cocktail parties but are more at ease in one-to-one breakfast meetings]

In which conditions do I feel at ease discussing topics with people I don't know? How can I reproduce these conditions?

Finally, here are some hints for actions that you can take to improve your networking:

- Draw a list of the people that are the most important for your current endeavor

- Prepare a schedule of contact making, aiming at contacting two people every day

- Make sure you follow your contact plan.

How can I show a minimum Kung Fu Master mindset and behavior?

If your score is that low, you probably consider yourself to have an almost obsessive-compulsive syndrome, lacking focus and follow-up.

First of all, let's do a reality check. Our experience is that often what people think of themselves is not what others observe. Ask, or get a trusted person, to ask people who know you well to assess whether this is true or not.

Let's assume that you confirm that the general view is effectively that you are always reacting to events and show a low focus and prioritization capability in your day-to-day activities. Here are some powerful questions we suggest you ponder. It is much more effective to discuss them with a third party that will challenge you to respond candidly:

- What is really preventing me from stopping myself from reacting to external events?

- In which conditions am I really overwhelmed by outside requests and events? What can I do to avoid being in these conditions? [some examples: shutting down alerts for emails, taking two hours at the beginning of each day with the door closed to strategize, always sleeping overnight on any important decision and then reviewing it in the morning]

- In which situations (which might include personal situations) did I feel more relaxed and in control of the overall situation, having a high-level view and strategizing? How can I reproduce these situations?

Finally, here are some hints for actions that may help you to improve your focus and prioritization skills:

- Discuss with a close colleague a list of priority tasks, after having identified and verbalized the expected outcome of the project;

- Announce publicly your priorities so that everybody expects these to be your areas of focus;

- Create space for relaxation and unplugging to avoid being constantly in a reactive mode; have someone create a filter to avoid excessive disturbance.

How can I show a minimum Entrepreneur mindset and behavior?

If your score is that low, you probably consider that you are short-term focused and you have difficulties looking at the "big picture."

First of all, let's do a reality check. Our experience is that often what people think of themselves is not what others observe. Ask, or get a trusted person, to ask people that know you well to assess whether this is true or not.

Let's assume that this confirms that you are excessively focused on the short term and seem to be generally unable to anticipate long-term consequences of your actions. Here are some powerful questions we suggest you ponder. It is much more effective to discuss them with a third party who will challenge you to respond candidly:

- What is really preventing me from investing in the future and looking at the long term consequences?
- In which particular conditions am I really focused only on the short term? What can I do to avoid being in these conditions? [example: avoid reactive mode by protecting my calendar, systematically take the time to think about the long term consequences, keep a share of resources for long term investment]
- In which situations (which might include personal situations) did I manage to take sensible decisions that involved investment in the long term? How can I reproduce these situations?

Finally, here are some hints for actions:

- For any decision, draw a table showing short-term / long-term advantages / drawbacks
- Identify what is really the final outcome you are seeking and align your decisions with it, every time
- Set aside a fund or resources for investments in the long term

How can I show a minimum Team Coach mindset and behavior?

If your score is that low, you consider that you have a very low capability of operating a team and letting it express all its potential. You have experienced many decaying teams.

First of all, let's do a reality check. Our experience is that often what people think of themselves is not what others observe. Ask, or get a trusted person, to ask people that know you well to assess whether this is true or not.

Let's assume that your team coach capabilities are confirmed to be low. Here are some powerful questions we suggest you ponder. It is much more effective to discuss them with a third party that will challenge you to respond candidly:

- What is really preventing me from getting the team to express its talents? What is preventing me from having those difficult conversations with the troublemakers and making the decisions that are needed?

- In which conditions am I really overwhelmed by a team's dynamics to a point where I freeze? What can I do to avoid being in these conditions? [example: avoiding confrontational situations where you are alone, spending more time developing emotional connection with people]

- In which situations (which might include personal situations) did I feel more relaxed, part of a great team, even leading a great team? How can I reproduce these situations?

Finally, here are some hints for actions that might help you:

- Identify whether there are troublemakers in your team who present more of a nuisance than a benefit; if so, organize the appropriate tough conversations with someone at your side to support you;

- Identify the working style you expect from the team and let them know what your expectations are;

- Exchange with the team how you can work together to improve your effectiveness.

How can I show a minimum People Catalyst mindset and behavior?

If your score is that low, you believe that you are not good at developing people and helping them release their talents.

First of all, let's do a reality check. Our experience is that often what people think of themselves is not what others observe. Ask, or get a trusted person, to ask people that know you well to assess whether this is true or not.

Let's assume that people confirm that you are not effective at supporting people in their growth and talents. Here are some powerful questions we suggest you ponder. It is much more effective to discuss them with a third party that will challenge you to respond candidly:

- What is really preventing me from getting to know people more personally?

- In which conditions am I really avoiding any consideration of the particular capabilities of people? What can I do to avoid being in these conditions? [example: avoid being under pressure, ask for suggestions about who would be more suited for a job before deciding]

- In which situations (which might include personal situations) did I feel more relaxed and in what situations did I take the time to know the people around me and to build on their talents? How can I reproduce these situations?

Finally, here are some hints for actions that may help you to improve:

- Take the time to have one-on-one discussions with your extended team to talk about their family, private life, hobbies, etc.;

- In assignment decisions, if the assignment is urgent and critical, pick an experienced person and provide an understudy for developmental purposes; if the assignment is not as critical, pick a person for which it is a developmental experience and get him mentored by an experienced staff member;

- Support people taking the initiative to change their career path and offer them opportunities.

Your role compensation plan

Note: the Workbook referred to in this section is available to download for free at www.ProjectSoftPower.com. This Project Soft Power Workbook gives a template you can use to fill in all the written exercises.

You have now identified your strong areas, which you will develop further. You may have possibly identified a very weak area that you need to bring up to a minimum level. That's great progress. All that remains now is how to compensate for those areas which are just okay, and on which you don't necessarily want to spend too much time and energy developing, because there are people around you that are extremely good at it.

The following discussion will give you a very useful guide as to the diversity of the team that needs to surround you so that together, you achieve great results.

As a start, let's associate each of the five Project Soft Power roles with one person who is particularly brilliant at this. It is quite rare that one person can be extremely good at two roles; still, this could happen. In that case, you can put the name of the person twice. And remember, you are one of those persons, in your area of strength.

Written exercise (refer to the Workbook): For each of the Project Soft Power roles, write who in your team has a particular strength in this role.

Written exercise (refer to the Workbook): If there is a gap, a name missing in front of one role, think of who you could bring in your team to fill the gap (refer to the Workbook).

Explain the Project Soft power framework to your team so that each key person understands the role he or she plays in the overall picture.

Define your action plan

Building on these principles—developing your strengths further, compensating for your weaknesses, we have provided in the next pages a framework for you to devise an action plan to improve your effectiveness as a Project Soft Power leader. The Workbook available for download on www.ProjectSoftPower.com gives an available template for you to use.

Now, how you will really implement your action plan

Most personal development books stop at this stage. Because we want you to really implement your action plan and become the successful project leader you deserve to be, we need to go one step further.

You are now highly energized and motivated to further develop yourself. In particular, because we suggest that you develop your strengths, the areas you are good at and on which you have already got good feedback.

Still, there are some ways to accelerate your learning and development curve, and also, to make sure that you will indeed make the effort. These principles are proven by extensive research:

- To accelerate your learning, you need to make sure to practice these habits and skills as often as possible in situations where you get immediate candid feedback on your performance;
- To make sure that you will indeed implement your action plan, you need to make close connections aware of your commitment so that they will energize you and remind you of your commitment.

As you can see, both those principles are based on the involvement of a limited number of close connections. Only people you feel well connected with will give you the candid feedback you seek. And you will probably only feel comfortable sharing your development plan with those people.

This leads us to the following written exercises.

Written exercise (refer to the Workbook): Who are the people you can count on to give you candid feedback if you ask them, in the workplace? (At this stage, identify as many as possible, we'll filter them out later.) Identify those people even if they are not directly working with you at the moment; you can always devise a way that they can come and see you in action to give the feedback you need. Don't retain people unless it is totally impossible that they could see you in action on the workplace in one way or another.

Written exercise (refer to the Workbook): Who are the people you can count on to make sure you follow through on your commitment, if you ask them to do so, and with whom you are comfortable sharing your plan to improve your Project Soft Power skills? At this stage, identify as many as possible; we'll filter them out later.

Consider now these two lists and sieve down now who is going to be seated in your personal Change Council—it is an honor for them to serve your transformation into a successful project leader. Here are some guides:

- Your Change Council should be limited in size to 3-4 persons;
- Make sure you will really get candid feedback from the person if you ask; you need to be close enough with that person that this can happen.
- Your Change Council can include people external to the organization like coaches, consultants, or others, as long as they can effectively observe you in action to found their feedback on actual observations;
- For each person, identify what particular aspect he/she will be bringing you useful feedback and information on.

Now you just need to go and ask those people you have carefully picked to be part of your personal Change Council! That might be emotionally a bit demanding, but see it as asking for a favor. Plan for this request: plan as to how you are going to request it, and to ensure your personal emotional commitment, put down a date when you will make this request.

Written exercise (refer to the Workbook): Filter out the few people you will rely on in the implementation of your plan, and when/how will you advise them.

Well done! You are now on your way to become the successful project leader you deserve to be.

How Project Soft Power is Rooted in Emotional Work

In this last section of the book, we will delve more generally into the value of emotional work in our world today, how it supports Project Soft Power, and how to develop the right emotional competencies. It is the most fundamental section of the book, a section that will give you a deep insight into the mechanisms of success in today's world. The content might appear less directly applicable than the rest of the book and I am aware that some readers might find it less appealing. That's why I have kept it short. If you have made it to here, I strongly recommend you make the effort to read these next few pages.

Because our professional value today lies in the quality of our emotional work, it is vital for each of us to understand this issue. And it is important to reach emotional work mastery to become a successful leader. It is vitally important for our project leadership. In the world we live in, it has also broader applications in our lives in general.

If you fully understand the value of emotional work and know how to excel in it, surely you will be the leader that everybody seeks.

What is emotional work?

Emotional work is the voluntary application of emotional intelligence, both to ourselves and to others whom we interact with. There are three different levels of emotional work:

- The unconscious (or innate) level, or our natural way of managing ourselves emotionally, and of connecting with others at the emotional level;
- The conscious level, where we are aware of how emotions influence ourselves and our relationship with others;
- The working level, where beyond analyzing the situation, we consciously act at the emotional level, should it be on ourselves or as part of our relationship with others.

We use the term 'work' because it is an effort which requires application and experience. Like all work forms, by regular application and exercise, you can significantly improve your effectiveness at doing emotional work. And you can reach emotional work mastery. This requires a pattern of exercise, feedback, and effort, and it is eminently worthwhile in today's world.

How and why our own value is rooted in our emotional work today, and why it is a significant change from a few decades ago

The world has changed dramatically in the last few decades. Remember that in the second half of the twentieth century, we were living in a world of scarcity. We were buying whatever was available without much discussion because there was not much of it (think about cars, for example). In this classical Industrial Age, the most respected form of intelligence was intellectual intelligence, or processing capability. It very much determined your status (measured by grades at school and reflected by your diploma). As long-distance communication was scarce, the typical organization was hierarchical and pyramidal. Analysis capability of large volumes of data was a highly valuable capability, because it was required at the top of the pyramid to be able to make meaningful decisions based on the convergence of all the information from the lower levels of the organization.

The world has changed fundamentally, and we generally underestimate how powerfully this change will shape our life in the years to come. We are now living in a world of abundance. Long-distance communication is plentiful and very cheap. Organizations have become flat and network-based. They increasingly organize around temporary projects (refer to the Fourth Revolution book[1] for more details). In this new world, data processing capability is not so highly valued—computers do that much better. What is increasingly valued is the capability to do emotional work. Leaders are increasingly defined by their emotional work capability and less and less by their data-processing intellectual capabilities.

Successful people in all walks of life need to master emotional work. It is even more important in the case of a project leader. Being a temporary endeavor, emotional connection with the team and within the team needs to be established quickly and effectively to deliver the project. Yes, this might surprise you: you can learn to master emotional work.

What type of emotional work does the Project Soft Power Leader practice?

Emotional Work and Network weaving

There is a lot of emotional work involved in building and sustaining a network at a level deep enough to be useful (thus, much deeper than cocktail party networking). This is often overlooked. Yet it comes as a blatant reality when someone tries, for example, to build a network on today's virtual social networks.

The main emotional work areas are: the ability to listen, the ability to give freely without expecting a return, the ability to accept criticism and direct feedback, the ability to face reality. Let's dwell quickly on these competencies.

- The ability to listen is required to ensure a quick connection. Listening here means *active* listening: listening in a way where you are completely focused on the other person and the rest of the world does not count. We all can do it: that's how we listen when we are in love! We need to develop this competency so that we can apply it voluntarily when we need it. It is the key to a quick connection. Taking time to appreciate and to be empathic with others requires overcoming the focus on self which is too often driving our behavior;
- The ability to give freely without expecting a return is key to developing and enhancing a connection once it has been created by listening. Why is it so hard emotionally? Because we are giving something valuable without being certain of a reward which might happen in the future, thus feeling that we are getting poorer. Beyond this emotional self-defense mechanism, most wise people indicate that you will get back much more than you gave, and in unexpected ways. Without going into this philosophical debate, it is pragmatically what you

need to do to deepen your connections. Overcome your first feeling of impoverishment. Give value until you think you've given too much, and then give a bit more.

- The ability to accept constructive criticism and direct feedback. This is a hard one and many people move back when criticized, and act to avoid this situation. In an office environment, we come across direct criticism and feedback quite rarely (although unfortunately it may abound behind one's back if you don't show that you accept it constructively). That's why every time it reaches you it is a gift to be looked at appreciatively even if you don't understand what it means on the spot. Just say thank you and consider it (you are not obliged to follow it). You need to reach a point where you welcome the most destructive criticism and feedback. That's emotionally difficult because each piece of criticism feels like an attack on our own identity. Overcome it. What is the worst that can happen? You won't die because of it.

- The ability to maintain a connection even in front of an abusive stakeholder is another skill that involves a lot of self-control and emotional stability. In front of abuse we tend to retreat into our shell, and our self-esteem feels attacked. Emotional stability is key in these instances, and can be best kept if the project leader boasts strong secure bases in his personal life, as well a very clear personal purpose that will allow him to put the situation back into perspective.

- Finally, it might surprise you that the ability to face reality is an emotional competency. It is fully related to the acceptance of feedback. In addition, it requires us to seek information that might contradict what we think or hope happens. For a project leader, it requires him to go on site and seek out and experience the reality. Many average project managers shun this under the pretext of overwork, but in fact it is because they are scared to discover something which is not in line with their expectations.

Emotional work, Discipline and Focus

The two main emotional components of the Kung Fu Master role are focus and discipline.

The first emotional component is focus, prioritization. As we have seen, it mainly revolves about the ability to say "no," to let go and to not do those things which are not a priority. It is difficult for us to stop doing things, just as it is difficult to get rid of what we have accumulated in our attic although we don't need it! The lingering thought of whether one of these things that we don't do will not come back with a vengeance or the scare of missing things and opportunities will stop us from dropping activities, and effectively focus on real priorities. Letting go is a key emotional competency identified by the Buddhist thinkers on the way to happiness. Overcome this emotional stoppage by remembering the 80/20 rule and defining priorities through a process which is shared with others. By having a traceability of why things have been prioritized and some public consent, you'll be able to overcome this block and fully focus.

A note on focus in today's world is needed here. It is the same mechanism at work when we take all the calls we get on our mobile whatever the circumstance, or even worse, when we set our computer or phone to beep whenever we get an email. Our world today promotes interruptions and thus undermines focus on our activities. Most of these interruptions are not important. Prioritize by dropping them and making sure you have adequate spans of time where you can really focus on what you have to do.

When it comes to discipline, most people fail. Most people do not maintain rigorous diet or exercise, or procrastinate fully when it comes to starting some difficult endeavor involving a large effort upfront. The emotional mechanism at work is linked to delayed gratification. Why do an effort today (with no visible benefit today) for a possible benefit in the distant future? We relinquish effort and suffering that does not have an immediate benefit.

There is a whole raft of literature on the subject, which focuses in particular, on education and how we should teach youngsters that benefits often require a significant dose of effort up front. Still, success is generally a question of doing a major effort upfront and 'overnight success' is just the manifestation of the realization of something that has been polished by years of disciplined practice. The only way to overcome this emotional issue is to focus on the expected purpose or benefit. Thus motivated by this vision of the future, we will endure the suffering of today and the grind of tomorrow and go the extra mile in our practice.

Emotional work and Entrepreneurial mindset

The emotional mechanism of an entrepreneurial mindset is one of delayed gratification, the same emotional mechanism as discipline. It is a matter of investing today for a benefit that might happen in the future. In addition, there is often a healthy dose of fear, as the outcome of any entrepreneurial venture is often uncertain. This particular fear is related to our identity, the risk of looking ridiculous or being criticized by the crowd for having made this investment.

As for discipline, to overcome these emotional blockages, the recipe is to focus on the ultimate purpose, and to share the decision-making process with a group.

Focusing on a strong purpose is also important to being able to put the inevitable intermediate setbacks into perspective and to maintaining an optimistic outlook, which is so important for the team's energy. Even when the situation is difficult, the successful leader conveys confidence in the outcome. It is not just a role the leader plays; it is because he looks further into the future than others do and knows that the path will be difficult, yet the team will still get there.

Emotional work and team coaching

In the Team Coach and People Catalyst roles, the emotional competencies lie more in the realm of influencing others' emotions. Of course, as a basis, it still requires a good level of self-mastery delivered by the emotional competencies described in the previous roles. We will not repeat them here, and instead will focus specifically on inter-personal emotional competencies.

In the Team Coach role, the successful project leader effectively creates a strong emotional connection between the team members. This requires that the team merges into one single team identity, above the identity, self-interest and ego of the team members. When you add that there needs to be a strong diversity in the team to enhance creative conflict between different points of view, the difficulty of the task could appear overwhelming.

The only way to achieve this level of emotional connection is to share a common purpose that enthuses the team and allows them to temporarily subordinate their own identity to the project identity. It is easier when it is a milestone or groundbreaking project that will be recognized as such by the industry; in all cases it is the art of the successful project leader to identify and communicate this superior project purpose, and use it as a basis to create a deep emotional connection.

To achieve emotional connection within the team, the successful project leader can start by developing strong emotional connections between him and the core members of the team, acting as a catalyst for the establishment of the same level of connection between the team members themselves. Then, it is important to bring in a little bit of vulnerability between the team members. Have the team members talk about their childhood, their fears, their hobbies, their families or any other subject that will open a slight crack in their usual social armor and allow the others to connect at a personal level. This needs to be done carefully, but it needs to get done. It is the main principle behind all the teambuilding activities: bring vulnerabilities of people to the foreground so that the others on the team can connect deeper.

Once this team emotional connection is in place, behaviors need to happen like positive, creative conflict and having the roles of each person evolve to fill in the gaps and take their talents more readily into account. Creative conflict requires that the purpose of the team be present above the egos of individuals, as well as an excellent emotional connection level. The evolution of roles beyond the title and the original job is even harder, because it means that individuals move away from their normal professional identity to take on a hybrid role, often without a specific denomination, that is more suited to them and the purpose of the team.

The team coach needs to constantly monitor and avoid having individual identities and egos strengthen and overcome the team identity. This is why it is important that the team coach prunes out of the team any individual which refuses to fit in (for example, because of a too strong ego that he does not want to subordinate to the team identity). More generally, this is why it is vital that the team coach can have tough conversations with team members if needed. It is also why the team emotional connection needs to be maintained and strengthened throughout the project by reminding people of the upper purpose of the project and developing further the project identity in itself.

Emotional work and catalyzing people's development

Real tough conversations are emotionally loaded. They are very different from screaming at people, which is a common way people avoid having the tough conversation! Tough conversations require a very high level of emotional engagement from both parties. Giving candid feedback in a useful way requires a lot of emotional self-mastery from the person giving feedback as well as from the person receiving it. How often have you held off from giving feedback for fear of provoking a damaging reaction? Developing a tight conversation after this feedback requires the ability to maintain a tight emotional connection with the individual, going possibly very deep into the person's emotional roots.

Our observation as coaches is that most people do actively avoid tough conversations involving negative feedback, thus creating situations that rot and can be destructive for the organization for a long time to come. How often we observe that people that are retrenched are surprised of what is happening to them because nobody had ever dared to tell them that there was a problem! The fear of conflict prevents people from acting. Still, there are simple techniques one can use to give feedback to people that can be used to defuse the conversation, and make the moment professional (see frame 1).

Frame 1. A powerful technique for effective feedback

Step 1: Be present *(environment, moment, focus)*
Step 2: Ask permission to give feedback *(giving permission is a powerful psychological opening)*
Step 3: State your purpose and <u>positive</u> intention
Step 4: Share personal perception of performance or behavior *(be specific, own the observation)*
Step 5: Provide suggestions for improvement / change
Step 6: Ask open question to get remarks
Step 7: Listen attentively
Step 8: Decide action(s)

As a People Catalyst, you will have to deal with the person's own professional identity. Often, while people know they have talents and know the areas they are passionate about, the step needed to bring these talents into play in a professional environment is very difficult. It is because their professional identity has been defined following some social convention and it seems dangerous to them to change it. If they exit or even blur their current professional identity, they will have to build an unconventional identity which will be criticized and maybe not recognized. Gentle support from the entire team, and subordination of one's identity to the team's purpose, is the engine of the transformation.

Some people will only change their professional identity temporarily during the project, and afterward, will take back their previous one; others will take advantage of the project to change significantly their professional identity and their life. Both options are very much respectable; what is important is that during the life of the project, people don't get stuck in their theoretical professional identity and use that as an excuse not to contribute their talents to the team.

The first steps people take outside their usual professional comfort zone will be scary and difficult. The support from the team and the leader is crucial, based on a deep emotional connection.

Reaching Project Soft Power mastery

We have now listed the main emotional competencies that are in the background of the Project Soft Power framework. The good news is, anybody can learn and develop these competencies through practice. The bad news is, it takes a lot of effort and exercise to master them, exactly like it takes a lot of effort and practice to get perfect muscles (like dream abdominals). Because in today's world these emotional competencies are key to your success and your value as a professional, it is important to understand them and make the effort to develop them further.

Many of these practices are not only applicable at the work place; they are also important at home and in the other facets of our life.

In the previous Section you have developed a plan for developing your Project Soft Power strengths, recover from your more pressing weaknesses, and otherwise compensate for other areas where you are less comfortable. In reality, by applying this action plan, you will actually practice and develop your emotional competencies.

Beyond all the emotional competencies that have been exposed here, three prerequisites appear to be driving a successful Project Soft Power application:

- The availability of a strong, compelling purpose for oneself, and for the project: this allows us to overcome many obstacles to change in ourselves and in others;
- Conversely, the capability to let go of what is less important or not aligned with the main purpose;
- A generous approach with a deep respect of other people as the exceptional beings they are.

When you implement your Project Soft Power action plan, and when you are stuck, reflect on the fact that:

- What prevents you from acting is FEAR—False Evidence Appearing Real;

- This is all a human adventure, with flesh-and-blood beings; organizations and institutions will not replace the effectiveness of mastering one's emotions and the emotional connection between individuals.

While you implement your Project Soft Power action plan, enjoy the discovery of yourself!

Are you ready?

References

1.   *The Fourth Revolution, How to Thrive Through the World's Transformation*, Jeremie Averous, Fourth Revolution Publishing, 2011: a description of the current transformations of the world as we move from the Industrial Age into the Collaborative Age

Conclusion – Project Soft Power, the Ultimate Differentiator

That's it! You now have in-depth knowledge of what Project Soft Power is about; you understand how it is critical for you to master Project Soft Power to be a reliably successful project leader. And, after having identified your strengths and weaknesses amongst the five roles of the SPIDER, the KUNG FU MASTER, the

ENTREPRENEUR, the TEAM COACH and the PEOPLE CATALYST, you also have an action plan to practice and improve further your Project Soft Power capabilities.

Remember—this is not a quest you will do alone. You will rely on your team to compensate for those roles that are not your stronger points. You will rely on your network for feedback and support to develop and practice your key Project Soft Power strengths.

Through a consistent practice of Soft Power, you will become a successful project leader, and you will improve your life in general.

You will become much more effective than all those project managers who focus on technical issues and believe that project management is just a set of processes that have to be followed mechanically and by the book. You will differentiate yourself as a Master Project Leader, who will be given the most challenging and exciting endeavors.

Project Soft Power will also change you deeply in your personal life. The practice of the five roles will become part of yourself, and they will live through all your personal endeavors and relationships, making you a better person.

Project leadership is always a human adventure. It is about creating something that has not been done before. Through Project Soft Power, you will catalyze the passion of the team and overcome the reluctance of the stakeholders whose life will be changed by your project.

The Value you will create and unleash through Project Soft Power will be much greater than the pure delivery of the project objectives. You will free others to become more plainly what they are, and you will show people the way to unleash their potential to the world, creating tremendous value in the years to come after the end of the project.

With Project Soft Power you will be much more effective and able to change the world. At the end of the day, that's what our activities down on Earth are all about. Don't hesitate. You now know how to do it. Go for it.

Many Thanks

This book – and the Project Soft Power concept – would not have seen the light without the live demonstration of Project Soft Power during the Gumusut subsea project, a large, complex project executed by SapuraAcergy for Shell in Malaysia. Under the leadership of Babu Surendran and Franck Louvety, something unique has happened in this multicultural team. In a situation where the project team was left by itself to build from scratch its processes, infrastructure and organization, the team spirit has been such that all those who have been part of this adventure will remember it as a highlight of their professional career.

For me it was at the same time a fantastic part of my life; a time that allowed me to uncover and develop my passion in writing and teaching; and the springboard for my new career as a consultant, author and trainer.

It is definitely the hope that this book, and the application of Project Soft Power, will become more widespread. That more consistently projects are setup and run in a way that get the team give out its best and overcome seemingly impossible feats. And at the same time, that projects are lead in a way that allow team members to give out the best of their passion and capabilities.

I'd like to extend many thanks to those who have taken valuable time to review the manuscript of the book and made many excellent comments and suggestions: Babu Surendran, Franck Louvety, Gawain Langford, Vijay

Gopinathan, Anthony Nouveliere, Jean-Francois Penverne. To all, I look forward to continue to exchange ideas with you for a long time. Many other came to discuss during presentations, and gave valuable insights. To all of you, thanks for the ideas and the suggestions.

Finally, I'd like to thank my family for the patience and the trust ever renewed. Not only was I writing this book but I was also changing career and creating my own company, all at the same time. Thanks to Helene and the kids Emile, Paul and Charlotte for the understanding support through good, tense and tough.

Index

Your Notes

Project Value Delivery,

a Leading International Consultancy for Large, Complex Projects

This cutting-edge project management book is sponsored by Project Value Delivery, a leading international consultancy that **"Empowers Organizations to be Reliably Successful in Executing Large, Complex Projects"**.

Part of our mission is to identify and spread the world-class practices that define consistent success for project leadership. Ultimately, we want to be able to deliver a framework that makes Large, Complex Projects a reliable endeavor.

Project Soft Power is a crucial part of this framework as it is an indispensable skillset for leaders in projects.

Our approach to project success

At Project Value Delivery we believe that project success is based on three main pillars which require specific sets of skills and methodologies specific to Large, Complex projects. All three need to be strong to allow for ultimate success:

- Project Soft Power™ (the human side)
- Systems
- Processes

We focus on embedding these skills and methodologies in organizations through consulting, coaching and training appointments. We develop what organizations need and then help them implement it sustainably, transferring the knowledge and skills.

We recognize that to be effective, our interventions will involve access to confidential business information and make it a point to treat all information provided to us with the utmost confidentiality and integrity.

Our Products

Our products are directly related to our three pillars. We have developed proprietary methods and tools to deliver the results that are needed for Large, Complex projects. In a number of areas, they are significantly different from those conventional project management tools used for simpler projects.

We focus on consulting, coaching and training interventions where we come in for a short to medium duration, analyze the situation, develop customized tools if needed, and transfer skills and methods to our clients so that they can implement them in a sustainable manner.

Contact

Contact us to learn more:

Contact @ ProjectValueDelivery.com, and visit our website **www.ProjectValueDelivery.com** where you can register to receive regular updates on our White Papers.

We Empower Organizations to be Reliably Successful in Executing Large, Complex projects.
Discover more on
www.ProjectValueDelivery.com